Simple Theories and Hyperimaginaries

In the 1990s Byunghan Kim and Anand Pillay generalized stability, a major model-theoretic idea developed by Saharon Shelah twenty-five years earlier, to the study of simple theories. This book is an up-to-date introduction to simple theories and hyperimaginaries, with special attention to Lascar strong types and elimination of hyperimaginary problems. Assuming only knowledge of general model theory, the foundations of forking, stability, and simplicity are presented in full detail. The treatment of the topics is as general as possible, working with stable formulas and types and assuming stability of the theory only when necessary. The author offers an introduction to independence relations as well as a full account of canonical bases of types in stable and simple theories. In the last chapters, the notions of internality and analyzability are discussed and used to provide a self-contained proof of elimination of hyperimaginaries in supersimple theories.

Enrique Casanovas is a Professor of Logic and Philosophy of Science in the Department of Logic, History and Philosophy of Science at the University of Barcelona.

LECTURE NOTES IN LOGIC

A Publication of
The Association for Symbolic Logic

This series serves researchers, teachers, and students in the field of symbolic logic, broadly interpreted. The aim of the series is to bring publications to the logic community with the least possible delay and to provide rapid dissemination of the latest research. Scientific quality is the overriding criterion by which submissions are evaluated.

Editorial Board

H. Dugald Macpherson, Managing Editor
Department of Pure Mathematics, School of Mathematics, University of Leeds

Jeremy Avigad
Department of Philosophy, Carnegie Mellon University

Vladimir Kanovei
Institute for Information Transmission Problems, Moscow

Manuel Lerman
Department of Mathematics, University of Connecticut

Heinrich Wansing
Faculty of Philosophy and Educational Science, Ruhr-Universität Bochum

Thomas Wilke
Institut fur Informatik, Christian-Albrechts-Universität

More information, including a list of the books in the series, can be found at http://www.aslonline.org/books-lnl.html.

Simple Theories and Hyperimaginaries

ENRIQUE CASANOVAS
Universidad de Barcelona

ASSOCIATION FOR SYMBOLIC LOGIC

CAMBRIDGE
UNIVERSITY PRESS

CAMBRIDGE UNIVERSITY PRESS
Cambridge, New York, Melbourne, Madrid, Cape Town,
Singapore, São Paulo, Delhi, Tokyo, Mexico City

Cambridge University Press
32 Avenue of the Americas, New York, NY 10013-2473, USA

www.cambridge.org
Information on this title: www.cambridge.org/9780521119559

First published 2011

Printed in the United States of America

A catalog record for this publication is available from the British Library.

Library of Congress Cataloging in Publication data

Casanovas, Enrique, 1957–
Simple theories and hyperimaginaries / Enrique Casanovas.
p. cm. – (Lecture notes in logic)
Includes bibliographical references and index.
ISBN 978-0-521-11955-9
1. Model theory. 2. First-order logic. 3. Hyperspace. I. Title. II. Series.
QA9.7.C37 2011
511.3′4 – dc22 2011005729

ISBN 978-0-521-11955-9 Hardback

Para Maribel

CONTENTS

PREFACE

These lecture notes originated in a seminar on Model Theory that I gave in the academic years 2005–06 and 2006–07 at the Department of Logic, History and Philosophy of Science of the University of Barcelona. I had presented some previous work on the basic notions of simple theories in July 2002 in the *Simpleton Workshop* held at the Centre International de Rencontres Mathématiques, Luminy (Marseille), which was subsequently published as [8]. A more extended version, including the exposition of stable theories, was the topic of a tutorial entitled *Advanced Stability Theory* that I taught in the *Modnet Summer School* that took place at the University of Freiburg in April 2006. And in preparing the material, I also drew on some courses on these topics given at the Universidad de los Andes, Bogotá, in August 2000 and in August 2004.

The notes are based on the work of many model theorists. The names of John T. Baldwin, Ehud Hrushovski, Byunghan Kim, Daniel Lascar, Ludomir Newelski, Anand Pillay, Bruno Poizat, Saharon Shelah, Frank O. Wagner, and Martin Ziegler deserve special mention. I learned stability theory from Martin Ziegler and I have made as much use as I could of his short and elegant proofs, presented in his courses and in his unpublished lecture notes.

This book is not as ambitious as Frank O. Wagner's book on simple theories [41], but its pace might be more comfortable for the beginner. I have tried to fill in some gaps and give details of proofs, but this has meant that fewer topics can be covered. Even in the realm of pure simple theories there are many important areas that have not been discussed here. The aim was to present the foundations of simple theories, forking calculus, and the stable fragment of any theory with the greatest generality I could afford and, secondly, to develop the topics of Lascar strong types and hyperimaginaries exhaustively.

My treatment of hyperimaginaries is complete up to two particular results: the theorem of Lascar–Pillay, based on Weil's theorem on compact groups, stating that every bounded hyperimaginary is equivalent to a sequence of finitary hyperimaginaries, and the still unpublished result of Adler that in every theory with the strict order property there is an infinitary hyperimaginary that is not eliminable (see [26] for the first result and [1] for the second one). I have also omitted the discussion of hyperimaginaries in the context of abstract elementary classes and some recent results within the nonindependence property setting.

Elimination of hyperimaginaries plays a fundamental role in the book. Let me summarize the situation. Stable and supersimple theories eliminate hyperimaginaries, low theories eliminate bounded hyperimaginaries, and small theories eliminate finitary hyperimaginaries; theories with the strict order property do not eliminate infinitary hyperimaginaries. Moreover, there is an example in [11] of a nonsimple theory without the strict order property that does not eliminate finitary hyperimaginaries. This is all we know.

The ultimate goal is to lay out, with all its prerequisites, a fundamental theorem in Model Theory due to Buechler, Pillay, and Wagner: supersimple theories eliminate hyperimaginaries. The last two chapters of the book are devoted to presenting this result.

In the Preliminaries chapter I fix notation and explain the framework, including the monster model, basic results on type-definability, imaginaries, and a very powerful result on indiscernibles based on the Erdős–Rado Theorem. I assume the reader is familiar with Model Theory. As basic reference texts I recommend the books of Hodges [13], Marker [28], Poizat [36], the lecture notes of Ziegler, and the recent books of Lascar [24], and of Tent and Ziegler [40].

Chapters 2–5 are devoted to the foundations of simple theories. In chapter 2 φ-types are discussed and the notions of λ-stability, stability, order property, independence property, strict order property, tree property, and simplicity are introduced. In chapter 3 local D-rank and dividing are defined and their basic properties are developed. In chapter 4 forking and dividing are discussed. In chapter 5 the ternary relation of independence based on nonforking is presented and, after giving a list of its more general properties that hold in any theory, we confine our study to the context of simple theories, proving local character, extension, symmetry, and transitivity properties. Morley sequences are also introduced and special attention is given to type-definability issues of nonforking and Morley sequences.

In chapters 6–8 the stable fragment of any theory is studied. I adopt a local point of view, as in Pillay [32] and [33], focusing the study on the behavior of φ-types for stable formulas φ in arbitrary theories. There are two different notions of φ-formulas and φ-types in the literature. I reserve the denomination of *φ-formula* and *φ-type* for Shelah's notion and I use the terms *generalized φ-formula* and *generalized φ-type* for the variants introduced by Pillay. The treatment of local Cantor–Bendixson rank in chapter 6 is based on Ziegler's lecture notes [42]. This rank $\mathrm{CB}_\Delta(\pi(x))$ is not the same as Shelah's local rank $R(\pi(x), \Delta, \infty)$, unless $\pi(x)$ is a set of Δ-formulas. The presentation of the topic of heirs and coheirs in chapter 7 is not complete since some interesting results of Lascar and Pillay (see [25]) have not been included.

Chapter 9 is a general introduction to Lascar strong types, Kim–Pillay-types, and strong types. Newelski's Theorem on finiteness of diameter of type-definable Lascar strong types (Theorem 9.22 here) is proved following ideas from [30] as developed in Peláez's doctoral dissertation [31].

Chapter 10 is devoted to the Independence Theorem for simple theories. First I prove the Independence Theorem for Lascar strong types (as Shami does in [37]) and then I derive from it the version for types over models. We take occasion to define G-compactness and present a few results about it. However, the Lascar group is not discussed here.

In chapter 11 canonical bases of stationary types in simple theories are introduced. This is just the classical theory of canonical bases in stable theories, but slightly generalized to make it as well applicable in simple theories. The proper notion of canonical base for simple theories is postponed to chapter 17 since it requires the previous introduction of hyperimaginaries.

Chapter 12 is an exposition of abstract independence relations, a topic of interest beyond simple theories. The treatment is axiomatic with special attention to nonforking and nondividing independence. Results from Adler's two articles [2] and [3], such as the symmetry of any independence relation (Corollary 12.6 here) or the study of $\mathop{\smile\!\!\!\!\!\!|}\,^*$, play a fundamental role. The Kim–Pillay Theorem characterizing simple theories as theories with an independence relation satisfying the Independence Theorem over models is also proved, as well as similar characterizations of stable theories. Finally, Kim's characterization (from [18]) of simple theories in terms of symmetry or transitivity of nonforking independence is given.

Chapters 13 and 14 are an exposition of supersimple theories with special emphasis on ranks SU and D. There is also an axiomatic characterization of ranks useful for characterizing superstable theories. On this last point I follow Ziegler's lecture notes.

With chapter 15 begins the second part of the book, whose main point is to develop the model theory of hyperimaginaries. The main references are Hart, Kim, and Pillay's seminal paper [12], Wagner's book [41], and Lascar and Pillay's article [26]. The first topics in chapter 15 are the equivalence of hyperimaginaries, types of hyperimaginaries, and the bounded closure operator. Then I study equality of Kim–Pillay-type relativized to a complete type and I give the proof that any bounded hyperimaginary is an equivalence class of a bounded type-definable equivalence relation (Proposition 15.27 here).

In chapter 16 forking and independence are developed for hyperimaginaries. It is not a straightforward translation of the previously studied theory and very often the natural order of proving some results from previous ones changes dramatically. For instance, the local character of forking (Proposition 16.21

here) has a very different proof compared to the classical case. In a first step the Independence Theorem is proved for types over models and later for hyperimaginary Lascar strong types. We also pay attention to type-definability of nonforking and independence.

The main application of hyperimaginaries is to provide canonical bases for amalgamation bases in simple theories. This is presented in chapter 17. In the first section of this chapter several ways of obtaining amalgamation bases in simple theories are examined carefully.

Chapter 18 deals with the elimination of hyperimaginaries and related topics. After presenting some general results about elimination, the cases of stable theories, small theories and low theories are considered. The elimination of hyperimaginaries in supersimple theories is the topic of chapter 20. In chapter 19 all prerequisites for the elimination proof are developed in detail. This concerns mainly the study of the analysability rank R^{an}, which might be interesting in itself for further applications. Chapter 20 closes with Kim's characterization of canonical bases in simple theories as sets of canonical parameters of definitions of p-stable formulas. We follow the short proof given by Kim and Pillay in [21].

Examples and commentaries are scarce throughout the book. My main concern has been with the compact presentation of the theory and the detailed exposition of proofs. I wish I had had the time and talent to embellish the text with funny comments and jokes, as the excellent writer Bruno Poizat does.

Acknowledgments. I thank the participants in the Model Theory Seminar, Hans Adler, Silvia Barbina, Rafel Farré, Javier Moreno, Rodrigo Peláez, Juan Francisco Pons, and Joris Potier for their patience and their remarks. I thank Daniel Palacín and Joris Potier for their careful reading of previous versions. I also thank Anand Pillay and Frank Wagner for answering many questions I addressed to them concerning details of proofs of their results on the topics of this book. Finally, I thank an anonymous referee of the book for his detailed remarks and suggestions. Of course, any mistakes or inaccuracies that might still remain in the text are my sole responsibility.

Chapter 1

PRELIMINARIES

T is a complete theory of language L with infinite models and $\mathfrak{C}$ is its monster model. Usually A, B, C are subsets of $\mathfrak{C}$ and a, b, c are tuples of elements of $\mathfrak{C}$. We think of $\mathfrak{C}$ as a proper class, so *subset* means small subclass. Some definitions also make sense for arbitrary subclasses A of $\mathfrak{C}$. For instance, for any subclass A of $\mathfrak{C}$, $\text{tp}(a/A)$ is the class of all formulas with parameters in A realized by a in $\mathfrak{C}$, and $\text{Aut}(\mathfrak{C}/A)$ is the group of all automorphisms of $\mathfrak{C}$ which fix A pointwise. If A is a set, then $\text{tp}(a/A) = \text{tp}(b/A)$ if and only if $f(a) = b$ for some $f \in \text{Aut}(\mathfrak{C}/A)$. Notation $a \in A$ means that all the elements in the tuple a belong to A. We use x, y for single variables but also for tuples of variables. A tuple is a sequence, not necessarily finite. A class A of tuples is *bounded* if it is small compared with the monster model, hence if it is a set. The reader may feel more comfortable assuming the monster model is a saturated model of inaccessible cardinality κ; in this case he should replace the words *proper class* and *set* by *of cardinality* κ and *of cardinality* $< \kappa$ respectively.

Unless stated otherwise, κ, λ, μ are infinite cardinal numbers and α, β, γ are ordinal numbers. On is the class of all ordinal numbers. We use i, j, k for indexes, in most cases ordinal numbers. Sometimes it is convenient to add an end point ∞ to On. Hence we understand that $\infty = \sup_{\alpha \in On} \alpha$. The set of all mappings from I into J is denoted by J^I. Hence, κ^λ has two meanings: cardinal exponentiation and set of functions. The context will make clear which one is intended. For a sequence $(a_i : i < \alpha)$ and any $j \leq \alpha$ we use $a_{<j}$ for $(a_i : i < j)$. The union of two sets A, B is sometimes denoted by AB.

A formula is *over* A if it has parameters in A. By $\varphi(x) \in L(A)$ we mean that $\varphi(x)$ is over A. In particular $\varphi(x) \in L$ means that $\varphi(x)$ is a formula without parameters. A partial type over A is a consistent set of formulas over A. A *finitary type* is a type in finitely many variables. For a formula $\varphi(x)$ we write $\models \varphi(a)$ or $a \models \varphi(x)$ to mean $\mathfrak{C} \models \varphi(a)$, and similarly for a type. The set of all complete types over A of a fixed length of tuples is denoted by $S(A)$. This makes sense even when A is a proper subclass of $\mathfrak{C}$, but in this case the types are not necessarily realized in $\mathfrak{C}$. If we want to make clear that n or κ is the fixed length of the involved tuples we write $S_n(A)$ or $S_\kappa(A)$.

$S(A)$ is naturally a *boolean topological space*, that is, a compact Hausdorff totally disconnected topological space. A basis of clopen sets is given by $[\varphi] = \{p \in S(A) : \varphi \in p\}$. If Σ is a type over A, the corresponding closed set is $[\Sigma] = \{p \in S(A) : \Sigma \subseteq p\}$.

We write $a \equiv_A b$ to express equality of types over A: $\text{tp}(a/A) = \text{tp}(b/A)$. For sets A, B we sometimes use notations like $\text{tp}(A/C)$ and $A \equiv_C B$. An implicit enumeration of A (and a corresponding one for B) should be assumed. Hence $A \equiv_C B$ means that $a \equiv_C b$ for some tuples a and b enumerating A and B respectively.

Implication is denoted by $\Sigma \vdash \varphi$. If Σ is a set of formulas, it means that all realizations of Σ in the monster model also realize φ. If Σ is a complete type over A and $\varphi \in L(A)$, this is obviously equivalent to $\varphi \in \Sigma$. If $\pi(x)$ is a partial type, $\pi(\mathfrak{C})$ is the class of tuples (of the length of x) that satisfy π, and more generally, $\pi(A)$ is the set of all tuples $a \in A$ such that $\models \pi(a)$. Equivalence of partial types $\Sigma(x)$ and $\pi(x)$ (perhaps over different sets) means $\Sigma(\mathfrak{C}) = \pi(\mathfrak{C})$. We write $\Sigma(x) \equiv \pi(x)$ for it. Of course, it can be rephrased as $\Sigma \vdash \varphi$ for all $\varphi \in \pi$ and $\pi \vdash \varphi$ for all $\varphi \in \Sigma$.

A model M is always an elementary submodel of the monster model $\mathfrak{C}$. In fact, $\mathfrak{C}$ is the union $\bigcup_{i \in On} M_i$ of an elementary chain $(M_i : i \in On)$ of models, where each M_i is $|i|$-saturated.

Very often we say a proof or a step in a proof is *by compactness*. Sometimes this means that if $\Sigma \vdash \varphi$ then $\Sigma_0 \vdash \varphi$ for some finite $\Sigma_0 \subseteq \Sigma$, but sometimes it only means we are claiming some set of formulas is consistent and we are using the fact that the monster model $\mathfrak{C}$ is saturated: every type over a (small) subset is realized in $\mathfrak{C}$. Of course, the consistency of the involved set of formulas should be checked. When we say that a set of formulas $\Sigma(x)$ is consistent or inconsistent we really mean consistency or inconsistency relative to our theory T. If $\Sigma(x)$ is over A it should be understood that we mean relative to the theory $T(A)$ of the monster model expanded by constants for the elements of A. This amounts to being the same for Σ as being finitely satisfiable in the monster model (expanded by the parameters in A).

It is convenient to work with complete types over the monster model. We call them *global types*. It makes sense to consider the space of types $S(\mathfrak{C})$ although every global type $\mathfrak{p} \in S(\mathfrak{C})$ is in fact a proper class. It is similar to what happens with the group $\text{Aut}(\mathfrak{C})$. The background set theory needed to deal with this situation is a finite iteration of the Bernays–von Neumann–Gödel theory of classes (with choice), which is a conservative extension of *ZFC*.

We often consider properties of formulas and discuss them. In fact they are usually properties of a formula φ together with a separation of disjoint tuples of variables x, y including all the variables occurring in the formula. The same formula with a different choice of tuples might not have the property.

The reader should notice that when we say $\varphi(x, y)$ has a certain property we really mean that the formula has the property with respect to the displayed separation of variables x, y. When several tuples appear and we want to make clear which is the intended separation, we use notations like $\varphi(x; y, z)$ or $\varphi(x, y; z)$. A particular case of all this appears when we start with a formula $\varphi(x, y)$ and we want to consider the same formula but with the opposite separation of tuples of variables: y, x. In this case we use $\varphi^{-1}(y, x)$ for the new situation. Hence $\models \varphi(a, b) \Leftrightarrow \models \varphi^{-1}(b, a)$.

The group $\mathrm{Aut}(\mathfrak{C}/A)$ acts naturally on $\mathfrak{C}$ but also on the space of types $S(\mathfrak{C})$. The image of an object by some $f \in \mathrm{Aut}(\mathfrak{C}/A)$ is often called an *A-conjugate* of the object by f. If $\varphi(x, y) \in L$, the A-conjugate of $\varphi(x, a)$ by $f \in \mathrm{Aut}(\mathfrak{C}/A)$ is the formula $\varphi(x, a)^f = \varphi(x, f(a))$. Similarly for types. If $\mathfrak{p} \in S(\mathfrak{C})$, the A-conjugate of $\mathfrak{p}$ by $f \in \mathrm{Aut}(\mathfrak{C}/A)$ is the global type $\mathfrak{p}^f \in S(\mathfrak{C})$ all whose formulas are conjugate by f of formulas in $\mathfrak{p}$. Hence $\mathfrak{p}^f = \{\varphi(x)^f : \varphi(x) \in \mathfrak{p}\}$. Sometimes it is convenient to look at $\mathrm{Aut}(\mathfrak{C}/A)$ as a topological group. The natural topology on $\mathrm{Aut}(\mathfrak{C}/A)$ is the topology of pointwise convergence. A basis of open sets is given by $\{f : f(a) = b\}$ for any finite tuples $a, b \in \mathfrak{C}$.

An element $a \in \mathfrak{C}$ is *definable over* the set Aif its orbit under $\mathrm{Aut}(\mathfrak{C}/A)$ is a singleton and it is *algebraic over A*if the orbit is finite. For any set A, the set of all elements definable over A is the *definable closure* of A, denoted $\mathrm{dcl}(A)$, and the set of all elements algebraic over A is the *algebraic closure* of A, denoted $\mathrm{acl}(A)$. These definitions can be extended to the case of an arbitrary class $A \subseteq \mathfrak{C}$ by letting $\mathrm{dcl}(A)$ be the union of all $\mathrm{dcl}(B)$ where B ranges over subsets of A, and similarly for $\mathrm{acl}(A)$. Notice that dcl and acl are finitary closure operators on subclasses of the monster model, i.e., they are operators cl such that

1. $A \subseteq \mathrm{cl}(A)$.
2. If $A \subseteq B$, then $\mathrm{cl}(A) \subseteq \mathrm{cl}(B)$.
3. $\mathrm{cl}(\mathrm{cl}(A)) \subseteq \mathrm{cl}(A)$.
4. If $a \in \mathrm{cl}(A)$, then $a \in \mathrm{cl}(A_0)$ for some finite $A_0 \subseteq A$.

A partial type $\pi(x)$ over A is called *algebraic* (over A) if all realizations of π belong to $\mathrm{acl}(A)$.

DEFINITION 1.1. A relation R (a subclass of $\mathfrak{C}^I$ for some index set I) is *A-invariant* if it is preserved under automorphisms of $\mathfrak{C}$ fixing A pointwise, that is,

$$R(a) \Rightarrow R(f(a))$$

for all a, for all $f \in \mathrm{Aut}(\mathfrak{C}/A)$. This clearly implies $R(a) \Leftrightarrow R(f(a))$.

REMARK 1.2. *For every set A, every A-invariant relation R is definable by a disjunction of types over A, namely*: $R(a) \Leftrightarrow a \models \bigvee_{R(b)} \bigwedge \mathrm{tp}(b/A)$.

DEFINITION 1.3. A *type-definable over A*relation R is a relation for which there is a set of formulas $\pi(x)$ over A such that $\pi(\mathfrak{C}) = R$, that is, $R(a) \Leftrightarrow\ \models \pi(a)$ for all a. If $A = \emptyset$ we say that R is *0-type-definable*. As usual, if π is in fact a formula we talk of *definability over A* or *A-definability*. A *0-definable*relation is a relation definable without parameters.

LEMMA 1.4. 1. *If R is type-definable over A and it is B-invariant, then it is type-definable over B. The same is true for definability.*
2. *If R and the complement $\bar{R}$ of R are type-definable over A, then R is definable over A.*
3. *The class of all type-definable over A relations is closed under arbitrary intersections, finite unions, and quantification of any number of variables.*

PROOF. 1. Let $\pi(x)$ be a type over A defining R and consider the restriction map $f : S(AB) \to S(B)$ given by $f(p) = p \restriction B$. It is a continuous function from a compact space onto a Hausdorff space, and therefore it is closed. The image of the closed set $[\pi] = \{p \in S(AB) : \pi \subseteq p\}$ in f is closed and therefore it is of the form $[\Sigma]$ for some partial type Σ over B. By B-invariance it is easy to check that $\Sigma(\mathfrak{C}) = R$.

2. Let $\pi_1(x), \pi_2(x)$ be sets of formulas over A defining R and its complement respectively. Since $\pi_1(x) \cup \pi_2(x)$ is inconsistent, there is a finite conjunction $\varphi(x)$ of formulas of $\pi_1(x)$ which is inconsistent with $\pi_2(x)$. Clearly, $\varphi(x)$ defines R.

3. For instance, if $\Sigma(x, y)$ defines R and it is closed under finite conjunctions, then $\{\exists y \varphi(x, y) : \varphi(x, y) \in \Sigma\}$ defines $\{a : \exists y R(a, y)\}$. ⊣

DEFINITION 1.5. Let $(I, <)$ be a linearly ordered set and let $a = (a_i : i \in I)$ be a sequence of tuples a_i of the same length. We say that a is *A-indiscernible* or is *indiscernible over A*if for every $n < \omega$, whenever $i_0 < \cdots < i_n$ and $j_0 < \cdots < j_n$ are indexes in I, then $a_{i_0}, \dots, a_{i_n} \equiv_A a_{j_0}, \dots, a_{j_n}$.

The existence of indiscernible sequences is usually established using Ramsey's Theorem. Here it is convenient to introduce a more powerful method based on the Erdős–Rado Theorem.

PROPOSITION 1.6. *If $\kappa \geq |T| + |A|$, $\lambda = \beth_{(2^\kappa)^+}$, and $(a_i : i < \lambda)$ is a sequence of tuples a_i of the same length $\leq \kappa$, then there is an A-indiscernible sequence $(b_i : i < \omega)$ such that for each $n < \omega$ there are $i_0 < \cdots < i_n < \lambda$ such that $b_0, \dots, b_n \equiv_A a_{i_0}, \dots, a_{i_n}$.*

PROOF. $(b_i : i < \omega)$ is a realization of a type $p = p(x_i : i < \omega) = \bigcup_{n<\omega} p_n(x_1, \dots, x_n)$ where $p_n \in S_n(A)$ and $p_n \subseteq p_{n+1}$. The types p_n are constructed inductively together with a descending chain $(I_n : n < \omega)$ of subsets of $(2^\kappa)^+$ of cardinality $|I_n| = (2^\kappa)^+$ and a family $(X_i^n : i \in I_n)$ of subsets X_i^n of λ in such a way that

1. $X_i^{n+1} \subseteq X_i^n$ for all $i \in I_{n+1}$.

2. $|X_i^n| > \beth_{2^\kappa+\alpha}$ if i is the α-th element of I_n.
3. $(a_{j_1}, \dots, a_{j_n}) \models p_n$ for all $j_1 < \dots < j_n \in X_i^n$ for all $i \in I_n$.

We start with $I_0 = (2^\kappa)^+$, $X_i^0 = \lambda$, and $p_0 = \emptyset$. Let $(\xi_\alpha : \alpha < (2^\kappa)^+)$ enumerate increasingly I_n, and set $I'_n = \{\xi_{\alpha+n} : \alpha < (2^\kappa)^+\}$. Let $i \in I'_n$, say $i = \xi_{\alpha+n}$ where $\alpha < (2^\kappa)^+$. There are at most 2^κ types $\mathrm{tp}(a_{j_1}, \dots, a_{j_{n+1}}/A)$ for $j_1 < \dots < j_{n+1} \in X_i^n$. Since $|X_i^n| > \beth_{2^\kappa+\alpha+n} = \beth_n(\beth_{2^\kappa+\alpha})$, by the Erdős–Rado Theorem ($(\beth_n(\kappa))^+ \to (\kappa^+)^{n+1}_\kappa$) there is some $X_i^{n+1} \subseteq X_i^n$ and some $p^i_{n+1} \in S_{n+1}(A)$ such that $|X_i^n| > \beth_{2^\kappa+\alpha}$ and for all $j_1 < \dots < j_{n+1} \in X_i^{n+1}$, $(a_{j_1}, \dots, a_{j_{n+1}}) \models p^i_{n+1}$. Since $|I'_n| = (2^\kappa)^+$ we can choose $I_{n+1} \subseteq I'_n$ such that $|I_{n+1}| = (2^\kappa)^+$ and $p^i_{n+1} = p^{i'}_{n+1} = p_{n+1}$ for all $i, i' \in I_{n+1}$. ⊣

COROLLARY 1.7. 1. *If $(a_i : i < \lambda)$ is indiscernible over A, there is some model $M \supseteq A$ such that $(a_i : i < \lambda)$ is indiscernible over M.*
2. *If $(a_i : i < \lambda)$ is indiscernible over A, then it is also indiscernible over* $\mathrm{acl}(A)$.

PROOF. 1. Choose a model $M \supseteq A$ and extend the sequence to an A-indiscernible sequence $(a_i : i < \lambda')$ large enough to apply Proposition 1.6. This gives an M-indiscernible sequence $(b_i : i < \omega)$ such that $(b_i : i < \omega) \equiv_A (a_i : i < \omega)$. Now extend this sequence to an M-indiscernible sequence $(b_i : i < \lambda)$ and choose N such that $(a_i : i < \lambda)N \equiv_A (b_i : i < \lambda)M$. Then $(a_i : i < \lambda)$ is indiscernible over the model $N \supseteq A$.
2. By 1, since $\mathrm{acl}(A) \subseteq M$ for any model $M \supseteq A$. ⊣

The many-sorted structure M^{eq} is constructed from $M \models T$, adding a new sort M^n/E for each 0-definable equivalence relation E on a finite power M^n of M and extending the language L to L^{eq} adding a corresponding n-ary function symbol π_E for the projection map, the map $\pi_E : M^n \to M^n/E$ which sends each $a \in M^n$ to its equivalence class $\pi_E(a) = a_E$. We identify M with $M/=$, and we call it the *home sort*. Note that all symbols of L are in L^{eq} and the home sort M carries its original structure. In fact the construction should be described syntactically, starting with the formula defining E instead of dealing with E itself. It is a uniform procedure in every model of T and everything applies in particular to the monster model $\mathfrak{C}$ of T. The corresponding theory T^{eq} is, by definition, the complete theory of $\mathfrak{C}^{\mathrm{eq}}$. Elements of $\mathfrak{C}^{\mathrm{eq}}$ are called *imaginaries*. We call elements of $\mathfrak{C}$ *real elements*. Sometimes it is convenient to consider imaginaries in $T(A)$. They are equivalence classes of tuples in A-definable equivalence relations and they are called *A-imaginaries*.

We define all notions for the one-sorted theory T, but it is a very instructive exercise to adapt the definitions to the many-sorted case and to check that the corresponding notions have the same properties in T^{eq}. This is not always trivial. Every general result we prove for a theory T can also be applied to T^{eq}, only taking care of the right translation to the many sorted framework.

LEMMA 1.8. *For every formula* $\varphi(y, x_1^{E_1}, \dots, x_n^{E_n}) \in L^{\mathrm{eq}}$, *where* y *is a tuple of variables of the home sort and each* $x_i^{E_i}$ *is a variable of sort* E_i *(an equivalence relation on* m_i*-tuples), there is a formula* $\psi(y, y_1, \dots, y_n) \in L$, *where each* y_i *is a new* m_i*-tuple of variables, such that for all tuples* $a, a_1, \dots, a_n \in M$ *of the corresponding length,*

$$M^{\mathrm{eq}} \models \varphi(a, \pi_{E_1}(a_1), \dots, \pi_{E_n}(a_n)) \Leftrightarrow M \models \psi(a, a_1, \dots, a_n).$$

PROOF. In the first step we replace all quantifiers by quantification over the home sort using new variables. For example, we replace $\exists x_i^{E_i}$ by $\exists y_i$ and then we replace every occurrence of $x_i^{E_i}$ by $\pi_{E_i}(y_i)$. In the second step the projections π_{E_i} are eliminated: $\pi_{E_i}(x) = \pi_{E_i}(x')$ is replaced by $E_i(x, x')$. ⊣

Notations like $\mathrm{dcl}^{\mathrm{eq}}(A)$ and $\mathrm{acl}^{\mathrm{eq}}(A)$ have the obvious meaning: definable and algebraic closure computed in $\mathfrak{C}^{\mathrm{eq}}$.

PROPOSITION 1.9. 1. $\mathfrak{C}^{\mathrm{eq}}$ *is the monster model of* T^{eq}.
2. $M^{\mathrm{eq}} \preceq \mathfrak{C}^{\mathrm{eq}}$ *if we identify each equivalence class* a_E *in* M *with the corresponding equivalence class in* $\mathfrak{C}$.
3. *Every automorphism of* M *extends uniquely to an automorphism of* M^{eq} *and we can identify* $\mathrm{Aut}(M)$ *with* $\mathrm{Aut}(M^{\mathrm{eq}})$.
4. $M^{\mathrm{eq}} = \mathrm{dcl}^{\mathrm{eq}}(M)$.
5. $\mathrm{dcl}^{\mathrm{eq}}(A) \cap \mathfrak{C} = \mathrm{dcl}(A)$ *and* $\mathrm{acl}^{\mathrm{eq}}(A) \cap \mathfrak{C} = \mathrm{acl}(A)$ *for every* $A \subseteq \mathfrak{C}$.
6. *Every relation definable in* M^{eq} *with parameters is definable in* M^{eq} *with parameters of* M.
7. *Every* $R \subseteq M^n$ *definable in* M^{eq} *over* $A \subseteq M$ *is definable over* A *in* M.

PROOF. Easy exercise. ⊣

A *canonical parameter* of a definable relation R is an imaginary element c such that for all $f \in \mathrm{Aut}(\mathfrak{C})$, $f(R) = R$ if and only if $f(c) = c$. It is unique up to interdefinability and it can be constructed in $\mathfrak{C}^{\mathrm{eq}}$ as follows: take some $\varphi(x, y) \in L$ and some tuple of parameters $a \in \mathfrak{C}$ such that $\varphi(\mathfrak{C}, a) = R$, define an equivalence relation E by

$$E(b, d) \Leftrightarrow \varphi(\mathfrak{C}, b) = \varphi(\mathfrak{C}, d)$$

and finally put $c = a_E$.

LEMMA 1.10. *Every* A*-imaginary is interdefinable with some imaginary.*

PROOF. If $e = a_E$ is an A-imaginary, the canonical parameter e' of the equivalence class a_E considered as a definable relation is interdefinable with e. ⊣

An equivalence relation is called *finite*if it has only finitely many classes.

PROPOSITION 1.11. *The following are equivalent for any definable relation* $R \subseteq \mathfrak{C}^n$ *and any set* $A \subseteq \mathfrak{C}$:

1. R *is definable over any model* $M \supseteq A$.
2. R *has only finitely many* A*-conjugates.*

3. *R is a union of equivalence classes of some finite A-definable equivalence relation.*
4. *R is definable over* $\mathrm{acl}^{\mathrm{eq}}(A)$.

PROOF. $1 \Rightarrow 2$. Assume R has infinitely many conjugates over A and choose a model $M \supseteq A$, say of cardinality $\kappa \geq |T|$. By compactness, R has at least κ^+ conjugates over A and they are all definable over M, which is impossible.

$2 \Rightarrow 3$. Let $R_1, \ldots, R_m$ be a list of all A-conjugates of R. Define $E(x, y) \Leftrightarrow \bigwedge_{1 \leq i \leq m}(R_i(x) \leftrightarrow R_i(y))$. E is a finite A-definable equivalence relation and R is a union of E-classes.

$3 \Rightarrow 4$. Assume E is a finite A-definable equivalence relation and R is the union of the classes $[a_1]_E, \ldots, [a_m]_E$. By Lemma 1.10 each $[a_i]_E$ is interdefinable with some imaginary e_i. Then $e_i \in \mathrm{acl}^{\mathrm{eq}}(A)$ and R is definable over $e_1, \ldots, e_m$.

$4 \Rightarrow 1$. Clear since $\mathrm{acl}^{\mathrm{eq}}(A) \subseteq M^{\mathrm{eq}}$ if $M \supseteq A$. ⊣

Iteration of the process $T \mapsto T^{\mathrm{eq}}$ is useless since $(T^{\mathrm{eq}})^{\mathrm{eq}}$ can be identified with T^{eq}. In particular each definable relation in $\mathfrak{C}^{\mathrm{eq}}$ has a canonical parameter in $\mathfrak{C}^{\mathrm{eq}}$.

COROLLARY 1.12. *If A is a finite set of imaginaries, there exists an imaginary c such that for all $f \in \mathrm{Aut}(\mathfrak{C})$, f fixes setwise A if and only if $f(c) = c$.*

PROOF. Since each finite tuple of imaginaries is interdefinable with a single imaginary, we may assume that all imaginaries in A are of the same sort. Hence A is a definable relation in $\mathfrak{C}^{\mathrm{eq}}$ and has a canonical parameter $c \in \mathfrak{C}^{\mathrm{eq}}$. ⊣

The Cantor–Bendixson rank of points and closed sets in a topological boolean space X plays a key role in the presentation of stable theories and it is also relevant for small theories. We give here a brief summary of the results that will be needed later. In the applications sometimes X will be $S(\mathfrak{C})$ or $S_\varphi(\mathfrak{C})$, which is a proper class. What we present here is also valid for this situation, with the only possible exception of Proposition 1.18.

DEFINITION 1.13. Let X be a topological space. For any $A \subseteq X$, let A' be the set of all accumulation points of A. If A is closed, then A' is a closed subset of A. For any set A, the topological closure of A in X is $A \cup A'$. The *Cantor–Bendixson derivative* of the space X is defined inductively as follows:

1. $X^{(0)} = X$.
2. $X^{(\alpha+1)} = (X^{(\alpha)})'$.
3. $X^{(\alpha)} = \bigcap_{i<\alpha} X^{(i)}$ if α is a limit ordinal number.
4. $X^{(\infty)} = \bigcup_{\alpha \in On} X^{(\alpha)}$.

Notice that each $X^{(\alpha)}$ is closed and $X = X^{(0)} \supseteq \ldots X^{(\alpha)} \supseteq X^{(\alpha+1)} \supseteq \cdots \supseteq X^{(\infty)}$.

REMARK 1.14. *Let X be a compact topological space.*

1. *If α is a limit ordinal and $X^{(i)} \neq \emptyset$ for all $i < \alpha$, then $X^{(\alpha)} \neq \emptyset$.*

2. *If α is the largest ordinal such that $X^{(\alpha)} \neq \emptyset$, then $X^{(\alpha)}$ is finite.*

PROOF. By compactness of X. ⊣

DEFINITION 1.15. Let X be a compact topological space. The *Cantor–Bendixson rank of a point $a \in X$*, $\mathrm{CB}_X(a)$, is ∞ if $a \in X^{(\infty)}$ and otherwise it is the largest ordinal α such that $a \in X^{(\alpha)}$. If $A \subseteq X$ is a nonempty closed set, we set $\mathrm{CB}_X(A) = \infty$ if there is some point $a \in A$ with $\mathrm{CB}_X(a) = \infty$ and otherwise $\mathrm{CB}_X(A)$ is the maximal ordinal α for which there is some $a \in A$ with $\mathrm{CB}_X(a) = \alpha$. In the first case the *Cantor–Bendixson degree of A* is ∞ and in the second case it is the number $n < \omega$ of points $a \in X$ with $\mathrm{CB}_X(a) = \alpha$. If $A = \emptyset$ we agree that $\mathrm{CB}_X(A) = -1$ and its Cantor–Bendixson degree is 0. The space X is *scattered* if $X^{(\infty)} = \emptyset$, that is, if every $a \in X$ has ordinal Cantor–Bendixson rank $\mathrm{CB}_X(a) < \infty$.

Recall that a boolean space is a compact Hausdorff space which is totally disconnected, i.e., which has a basis of clopen sets.

PROPOSITION 1.16. *Let X be a boolean topological space and assume $A \subseteq X$ is closed. There is some clopen set $U \supseteq A$ with the same Cantor–Bendixson rank and degree as A. Hence,*

$$\mathrm{CB}_X(A) = \min\{\mathrm{CB}_X(U) : U \supseteq A \text{ is clopen}\}.$$

PROOF. This is clear if $A = \emptyset$ or $\mathrm{CB}_X(A) = \infty$. Assume $0 \leq \alpha = \mathrm{CB}_X(A) < \infty$. Then $A \cap X^{(\alpha+1)} = \emptyset$ and for each $a \in A$ there is some clopen set U_a such that $a \in U_a$ and $U_a \cap X^{(\alpha+1)} = \emptyset$. By compactness finitely many sets U_a cover A and their union is a clopen set $U \supseteq A$ such that $U \cap X^{(\alpha+1)} = \emptyset$. There are only finitely many points $a \in U \smallsetminus A$ with $\mathrm{CB}_X(a) = \alpha$ and for each one we can find a clopen set V_a such that $a \in V_a$ and $V_a \subseteq U \smallsetminus A$. If V is the union of these clopen sets V_a, then $U \smallsetminus V$ is a clopen set extending A with the same Cantor–Bendixson rank and degree as A. ⊣

PROPOSITION 1.17. *In any boolean topological space X, we can compute the Cantor–Bendixson rank of a clopen set $U \subseteq X$ according to the following rules:*

1. $\mathrm{CB}_X(U) \geq 0$ *if and only if* $U \neq \emptyset$.
2. $\mathrm{CB}_X(U) \geq \alpha + 1$ *if and only if there is a sequence* $(U_n : n < \omega)$ *of pairwise disjoint clopen subsets* $U_n \subseteq U$ *such that* $\mathrm{CB}_X(U_n) \geq \alpha$ *for all* $n < \omega$.
3. $\mathrm{CB}_X(U) \geq \alpha$ *for a limit ordinal* α *if and only if* $\mathrm{CB}_X(U) \geq i$ *for all* $i < \alpha$.
4. *If* $\mathrm{CB}_X(U) = \alpha < \infty$, *then the Cantor–Bendixson degree of* U *is the largest* $n < \omega$ *for which there are pairwise disjoint clopen subsets* $U_1, \ldots, U_n$ *of* U *such that* $\mathrm{CB}_X(U_i) \geq \alpha$ *for all* $i = 1, \ldots, n$.

Moreover, the Cantor–Bendixson rank of a point $a \in X$ can be computed in terms of clopen sets as follows:

$$\mathrm{CB}_X(a) = \min\{\mathrm{CB}_X(U) : U \text{ is clopen and } a \in U \subseteq X\}.$$

PROOF. Left to the reader. ⊣

PROPOSITION 1.18. *Let X be a boolean topological space and assume $U \subseteq X$ is clopen. Then $\mathrm{CB}_X(U) = \infty$ if and only if there is a binary tree $(U_s : s \in 2^{<\omega})$ of nonempty clopen subsets $U_s \subseteq U$ such that U_s is the disjoint union of $U_{s\frown 0}$ and $U_{s\frown 1}$ for all $s \in 2^{<\omega}$.*

PROOF. $\Rightarrow$. Since X is a set, there is some ordinal α such that $X^{(\alpha)} = X^{(\alpha+1)}$. Then $\mathrm{CB}_X(a) = \infty$ if and only if $\mathrm{CB}_X(a) \geq \alpha$ for all points $a \in X$. Assuming $\mathrm{CB}_X(U) = \infty$, we start the construction of the tree with $U_\emptyset = U$. Assuming inductively $\mathrm{CB}_X(U_s) = \infty \geq \alpha + 1$ we can split U_s into two disjoint clopen sets $U_{s\frown 0}$ and $U_{s\frown 1}$ of rank $\geq \alpha$ and hence of rank $= \infty$.

$\Leftarrow$. Assume $0 \leq \mathrm{CB}_X(U) < \infty$ and choose U_s of minimal ordinal rank α and of minimal degree among the clopen sets in the tree of rank α. Since U_s is the disjoint union of $U_{s\frown 0}$ and $U_{s\frown 1}$, one of them must have smaller rank or the same rank and smaller degree, a contradiction. ⊣

COROLLARY 1.19. *If the boolean topological space X has smaller cardinality than the continuum 2^ω, then it is scattered, i.e., $\mathrm{CB}_X(a) < \infty$ for all $a \in X$.*

PROOF. By Proposition 1.18. ⊣

PROPOSITION 1.20. *If the boolean topological space X has a countable basis of open sets and it is scattered, then it is countable.*

PROOF. Fix a countable basis $\{O_n : n < \omega\}$ of clopen sets and choose for each point $p \in X$ with $\mathrm{CB}_X(p) = \alpha$ some $n < \omega$ such that O_n isolates p in $X^{(\alpha)}$. It is easy to see that the assignment is one-to-one. ⊣

Chapter 2

φ-TYPES, STABILITY AND SIMPLICITY

DEFINITION 2.1. Let $\varphi(x, y) \in L$. A *φ-formula* over A is a formula of the form $\varphi(x, a)$ or $\neg\varphi(x, a)$ with $a \in A$. A *φ-type over A* is a consistent set of φ-formulas over A. A *φ-type $p(x)$ over A* is *complete* if for every $a \in A$ either $\varphi(x, a) \in p$ or $\neg\varphi(x, a) \in p$. The set of all complete φ-types over A is $S_\varphi(A)$. If $p(x)$ is a complete type over A, the restriction $p \restriction \varphi$ is defined as the set of all φ-formulas of p. Clearly, $p \restriction \varphi \in S_\varphi(A)$. This terminology and notation also apply to global types.

REMARK 2.2. *One might also define φ-types over A as consistent sets of boolean combinations of φ-formulas over A. Every complete φ-type in our sense can be extended uniquely to a complete φ-type in this apparently wider sense. Hence, nothing new is really gained. But in the literature there is also a truly different notion of φ-formula over A. It is defined as a formula over A which is equivalent to a boolean combination of formulas of the form $\varphi(x, a)$ where a is not necessarily a tuple of A. See Definition* 6.9.

DEFINITION 2.3. Let $\varphi(x, y) \in L$. A complete φ-type $p(x)$ over B is *definable over A* or *A-definable* if there is a formula $\psi(y) \in L(A)$, called a *definition of p*, such that for all $a \in B$,

$$\varphi(x, a) \in p \Leftrightarrow \models \psi(a).$$

A standard notation for the defining formula $\psi(y)$ is $d_p x\varphi(x, y)$. If A is not mentioned, we understand that $A = B$. A complete type $p(x) \in S(B)$ is *definable* (over A) if all its restrictions $p \restriction \varphi$ are definable (over A).

LEMMA 2.4. *Let $p(x) \in S_\varphi(M)$ be definable and let $B \supseteq M$.*

1. *There is some $q(x) \in S_\varphi(B)$ extending p which is definable over M.*
2. *If $q_1(x), q_2(x) \in S_\varphi(B)$ are M-definable extensions of p, then $q_1 = q_2$.*

PROOF. 1. Let $\psi(y) \in L(M)$ be a definition of p and set

$$q(x) = \{\varphi(x, a) : a \in B, \models \psi(a)\} \cup \{\neg\varphi(x, a) : a \in B, \models \neg\psi(a)\}.$$

The completeness of q is clear, and its consistency follows from the fact that M (and hence also $\mathfrak{C}$) satisfies the sentences

$$\forall x_1 \dots x_n y_1 \dots y_n (\bigwedge_{1 \le i \le n} \psi(x_i) \wedge \neg\psi(y_i) \rightarrow \exists x (\bigwedge_{1 \le i \le n} \varphi(x, x_i) \wedge \neg\varphi(x, y_i))).$$

2. If $q_1, q_2 \in S_\varphi(B)$ are M-definable extensions of p with definitions $\psi_1(y), \psi_2(y) \in L(M)$, then $M \models \forall y(\psi_1(y) \leftrightarrow \psi_2(y))$, which implies $\mathfrak{C} \models \forall y(\psi_1(y) \leftrightarrow \psi_2(y))$ and therefore $q_1 = q_2$. ⊣

Definition 2.5. Let λ be an infinite cardinal number. We say that $\varphi = \varphi(x, y) \in L$ is *λ-stable* or *stable in λ* if for any set A,

$$|A| \le \lambda \Rightarrow |S_\varphi(A)| \le \lambda.$$

It is said that φ *is stable* if it is stable in some λ. Otherwise φ is called *unstable*.

Proposition 2.6. *The following conditions are equivalent for any formula $\varphi = \varphi(x, y) \in L$.*

1. *$\varphi(x, y)$ is stable.*
2. *$\Gamma_\varphi(\omega)$ is inconsistent, where for any ordinal α,*

$$\Gamma_\varphi(\alpha) = \{\varphi(x_f, y_{f \restriction i})^{f(i)} : f \in 2^\alpha, i < \alpha\}$$

and where $\varphi^0 = \varphi$ and $\varphi^1 = \neg\varphi$.

3. *For any set A, any type $p(x) \in S_\varphi(A)$ is definable.*
4. *$\varphi(x, y)$ is λ-stable for all λ.*

Moreover in 3. one can add that p is definable by a formula of the form

$$\psi(y) = \exists x_1 \dots x_n \exists y_1 \dots y_m \chi(y, x_1, \dots, x_n, y_1, \dots, y_m)$$

where χ is a conjunction of formulas of the form $\varphi(x_i, y_j), \neg\varphi(x_i, y_j), \varphi(x_i, y)$, and $\varphi(x_i, y)$-formulas over A.

Proof. $1 \Rightarrow 2$. Assume $\Gamma_\varphi(\omega)$ is consistent. Let λ be an infinite cardinal number and let μ be the least cardinal number such that $2^\mu > \lambda$. Then $2^{<\mu} \le \lambda$. Since $\Gamma_\varphi(\mu)$ is also consistent, there is a tree $(b_s : s \in 2^{<\mu})$ such that for every $f \in 2^\mu$ the set of φ-formulas $p_f(x) = \{\varphi(x, b_{f \restriction i})^{f(i)} : i < \mu\}$ is consistent. Since $p_f(x)$ is inconsistent with $p_{f'}(x)$ whenever $f \neq f'$, it follows that there are $2^\mu > \lambda$ complete φ-types over the set $A = \{b_s : s \in 2^{<\mu}\}$. Since $|A| \le \lambda$, this shows that $\varphi(x, y)$ is not λ-stable.

$2 \Rightarrow 3$. Let $p(x) \in S_\varphi(A)$ and assume $\Gamma_\varphi(\omega)$ is inconsistent. Then $\Gamma_\varphi(\omega) \cup \bigcup_{f \in 2^\omega} p(x_f)$ also is inconsistent and by compactness there is a least natural number n for which

$$\Gamma_\varphi(n) \cup \bigcup_{f \in 2^n} p(x_f)$$

is inconsistent. Again by compactness, there is a finite subset $p_0(x) \subseteq p(x)$ such that $\Gamma_\varphi(n) \cup \bigcup_{f \in 2^n} p_0(x_f)$ is inconsistent. Then $n > 0$ and one can check that for any $a \in A$,

$$\varphi(x, a) \in p \Leftrightarrow \Gamma_\varphi(n-1) \cup \bigcup_{f \in 2^{n-1}} p_0(x_f) \cup \{\varphi(x_f, a)\} \text{ is consistent,}$$

that is:

$\varphi(x, a) \in p$ if and only if

$$\models \exists (x_f : f \in 2^{n-1}) \exists (y_s : s \in 2^{<n-1}) (\bigwedge \Gamma_\varphi(n-1) \wedge \bigwedge_{f \in 2^{n-1}} p_0(x_f) \wedge \varphi(x_f, a))$$

which is a definition of p of the form indicated above.

$3 \Rightarrow 4$. Since there are at most λ many definitions of the described form over a set A with $|A| \leq \lambda$, there are also $\leq \lambda$ many complete φ-types over A. This shows that $\varphi(x, y)$ is λ-stable for any λ, but we are using not only hypothesis 3 but also the added information on the form of the definition. Without this extra information we can only guarantee that φ is stable in every $\lambda \geq |T|$. But in fact this is enough to prove $3 \Rightarrow 1$ and after all, using the additional information on the form of the definition, we have established that 1 implies 4. ⊣

REMARK 2.7. 1. *If $\lambda \geq |T|$ and all complete φ-types over models of cardinality λ are definable, then φ is λ-stable.*

2. *If φ is stable, then any global φ-type $\mathfrak{p} \in S_\varphi(\mathfrak{C})$ is definable.*

PROOF. 1. If $\lambda \geq |T|$, any parameter set A of cardinality $\leq \lambda$ can be extended to a model M of cardinality λ. Therefore $|S_\varphi(A)| \leq |S_\varphi(M)| \leq \lambda$, since there are at most λ definitions of φ-types over M.

2. The proof of $2 \Rightarrow 3$ given for Proposition 2.6 works also for $\mathfrak{p} \in S_\varphi(\mathfrak{C})$. ⊣

DEFINITION 2.8. $\varphi = \varphi(x, y) \in L$ has the *order property* if there are $(a_i : i < \omega)$, $(b_i : i < \omega)$, such that:

$$\models \varphi(a_i, b_j) \Leftrightarrow i < j.$$

REMARK 2.9. 1. *$\varphi(x, y)$ has the order property if and only if there are $(a_i : i < \omega)$, $(b_i : i < \omega)$, such that $\models \varphi(a_i, b_j) \Leftrightarrow i \leq j$.*

2. *In the definition of the order property one can change the index set ω and its order by any infinite linear ordering.*

3. *$\varphi(x, y)$ has the order property if and only if $\neg\varphi(x, y)$ has the order property.*

4. *$\varphi(x, y)$ has the order property if and only if $\varphi^{-1}(y, x)$ has the order property.*

PROOF. 1. Replace b_j by $b'_j = b_{j+1}$. For 2 apply compactness.

3. First use 2 with the reverse order ω^* of ω, and then replace the strict order by the corresponding reflexive one as in 1. Finally 4 follows from 2 using ω^* again. ⊣

LEMMA 2.10. *Assume* $\varphi = \varphi(x, y)$ *does not have the order property. If* $p(x) \in S_\varphi(A)$ *is finitely satisfiable in* A *(which is always true if* A *is a model), then* p *is definable by a positive boolean combination of formulas of the form* $\varphi(a, y)$ *with* $a \in A$.

PROOF. Let $X_1, \ldots, X_n$ be a family of subsets of a set A. Consider the relation $R(a, b)$ between elements a, b of A defined by

$$R(a, b) \Leftrightarrow \forall i(1 \leq i \leq n)(a \in X_i \rightarrow b \in X_i).$$

It is easy to see that a subset $X \subseteq A$ is a positive boolean combination of the sets $X_1, \ldots, X_n$ if and only if $b \in X$ whenever $a \in X$ and $R(a, b)$. The reason is that in this situation

$$X(x) \Leftrightarrow \bigvee_{a \in X} \bigwedge \{X_i(x) : a \in X_i\}.$$

We will use this result. Assume p is not definable by a positive boolean combination of formulas of the described form. We inductively choose tuples a_i, b_i, c_i $(i < \omega)$ of elements of A. Suppose a_j, b_j, c_j are defined for all $j < i$. By hypothesis, $\{a \in A : \varphi(x, a) \in p\}$ is not a positive boolean combination of the sets $X_j = \{a \in A :\models \varphi(c_j, a)\}$ for $j < i$. Then there are $a_i, b_i \in A$ such that $\varphi(x, a_i) \in p$, $\neg\varphi(x, b_i) \in p$ and for all $j < i$, if $\models \varphi(c_j, a_i)$, then $\models \varphi(c_j, b_i)$. Now let c_i be a realization in A of the finite type $p \upharpoonright \{a_j, b_j : j \leq i\}$. The sequences of tuples obtained this way have the property that $\models \varphi(c_j, a_i) \wedge \neg\varphi(c_j, b_i)$ for $i \leq j$ but $\models \varphi(c_j, a_i) \rightarrow \varphi(c_j, b_i)$ for $j < i$. By Ramsey's Theorem we may assume that always $\models \neg\varphi(c_j, a_i)$ for $j < i$ or always $\models \varphi(c_j, b_i)$ for $j < i$. In the first case we have $i \leq j$ if and only if $\models \varphi(c_j, a_i)$. In the second case $i \leq j$ if and only if $\models \neg\varphi(c_j, b_i)$. In either case $\varphi(x, y)$ has the order property. ⊣

PROPOSITION 2.11. $\varphi(x, y)$ *is stable if and only if it does not have the order property.*

PROOF. If $\varphi(x, y)$ has the order property, then there are $a_i, b_j (i, j \in \mathbb{Q})$ such that for all i, j

$$\models \varphi(a_i, b_j) \Leftrightarrow i < j.$$

Now for each real number r, let $p_r(x)$ be the φ-type $\{\varphi(x, b_j) : r < j\} \cup \{\neg\varphi(x, b_j) : r \geq j\}$. Clearly $p_r(x)$ is inconsistent with $p_s(x)$ if $r \neq s$ and thus there are 2^ω complete φ-types over the countable set $\{b_j : j \in \mathbb{Q}\}$. Hence φ is not stable. The other direction follows from Lemma 2.10 and the first point in Remark 2.7. ⊣

COROLLARY 2.12. *Any boolean combination* $\varphi(x_1, \ldots, x_n; y_1, \ldots, y_n)$ *of stable formulas* $\varphi_i(x_i, y_i)$ *is stable. The tuples* x_i *and* x_j *(and also* y_i *and* y_j*) may have elements in common, but* x_i *and* y_j *are assumed to be disjoint.*

PROOF. Without loss of generality $x_i = x_j = x$ and $y_i = y_j = y$. For the case $\neg\varphi(x, y)$ use Remark 2.9 and for $\varphi(x, y) \vee \psi(x, y)$ use Ramsey's Theorem. ⊣

REMARK 2.13. *Let $\varphi = \varphi(x, y) \in L$ be stable. By Remark 2.7, any $\mathfrak{p}(x) \in S_\varphi(\mathfrak{C})$ is definable over some set A. If $M \supseteq A$, then $\mathfrak{p}(x)$ has a definition which is a positive boolean combination of formulas of the form $\varphi(a, y)$ with $a \in M$ and which is, at the same time, equivalent to a formula over A.*

PROOF. Since φ does not have the order property, we can use Lemma 2.10, which gives a positive boolean combination $\psi(y) \in L(M)$ of formulas of the form $\varphi(a, y)$ which defines $\mathfrak{p} \restriction M$. Since there is only one global φ-type extending $\mathfrak{p} \restriction M$ and definable over M, and ψ defines in $\mathfrak{C}$ a φ-type with these properties, it follows that ψ defines $\mathfrak{p}$ and hence ψ is equivalent to a formula over A. ⊣

DEFINITION 2.14. $\varphi(x, y) \in L$ has the *strict order property* if there is a sequence $(a_i : i < \omega)$ such that for all $i < j < \omega$,

$$\varphi(\mathfrak{C}, a_i) \subsetneq \varphi(\mathfrak{C}, a_j).$$

REMARK 2.15. 1. *Clearly, a formula with the strict order property has the order property.*
2. *In the definition of the strict order property one can replace the ordered set $(\omega, <)$ by any other infinite linearly ordered set.*
3. *If the formula $\varphi(x, y, a)$ has the strict order property, then also $\varphi(x; y, z)$ has the strict order property.*
4. *There is a formula in T with the strict order property if and only if for some n there is a 0-definable partial order of $\mathfrak{C}^n$ which has infinite chains. In fact if $\varphi(x, y)$ has the strict order property, then*

$$\psi(y_1, y_2) = \forall x(\varphi(x, y_1) \to \varphi(x, y_2)) \wedge \exists x(\varphi(x, y_2) \wedge \neg\varphi(x, y_1))$$

defines such an order.

PROOF. It is an easy exercise. ⊣

DEFINITION 2.16. $\varphi(x, y) \in L$ has the *independence property* if there are sequences $(a_i : i < \omega)$ and $(b_X : X \subseteq \omega)$ such that for all i, X,

$$\models \varphi(a_i, b_X) \Leftrightarrow i \in X.$$

REMARK 2.17. 1. *A formula $\varphi(x, y)$ with the independence property is unstable. In fact, for every cardinal λ there is some set A of cardinality λ such that $|S_\varphi(A)| = 2^\lambda$.*
2. *$\varphi(x, y)$ has the independence property if and only if for each $n < \omega$ there is a sequence $(a_i : i < n)$ such that for each $X \subseteq n$*

$$\{\varphi(a_i, y) : i \in X\} \cup \{\neg\varphi(a_i, y) : i \in n \setminus X\}$$

is consistent.

3. *If $\varphi(x, y)$ has the independence property, then also $\varphi^{-1}(y, x)$ has the independence property.*

PROOF. 1 and 2 are compactness arguments. For 3, for any $n < \omega$, there are $(a_X : X \in \mathcal{P}(n))$, $(b_I : I \subseteq \mathcal{P}(n))$ such that $\models \varphi(a_X, b_I) \Leftrightarrow X \in I$. Let $U_i = \{X \subseteq n : i \in X\}$ for $i < n$ and let $c_i = b_{U_i}$. Then $\models \varphi^{-1}(c_i, a_X) \Leftrightarrow i \in X$. ⊣

PROPOSITION 2.18. *There is an unstable formula in T if and only if there is a formula with the strict order property or there is a formula with the independence property. In fact if $\varphi(x, y)$ is an unstable formula without the independence property, then the following holds:*

1. *Some conjunction of $\varphi(x, y)$ with formulas of the form $\varphi(x, a)$ and $\neg\varphi(x, a)$ has the strict order property.*
2. *For some $n < \omega$, for some $s \in 2^n$, the formula $\psi(x; y_1, \ldots, y_n) = \bigwedge_{i<n} \varphi(x, y_{i+1})^{s(i)}$ has the strict order property (where $\varphi^0 = \varphi$ and $\varphi^1 = \neg\varphi$).*

PROOF. As already remarked, formulas with the independence property or the strict order property are unstable. Now assume that $\varphi(x, y)$ has the order property but not the independence property. We will see that some conjunction $\theta(x, y)$ of $\varphi(x, y)$ with formulas of the form $\varphi(x, a)$ and $\neg\varphi(x, a)$ has the strict order property. By point 3 in Remark 2.15, this will suffice.

By the order property, there are sequences $(a_i : i \in \mathbb{Q})$ and $(b_i : i \in \mathbb{Q})$ such that $\models \varphi(a_i, b_j)$ if and only if $i < j$. We may assume that $(b_i : i \in \mathbb{Q})$ is indiscernible. Since φ^{-1} does not have the independence property, for some $n < \omega$ there is a subset $S \subseteq n$ which is not represented, in the sense that

$$\models \neg\exists x(\bigwedge_{i \in S} \varphi(x, b_i) \wedge \bigwedge_{i \in n \setminus S} \neg\varphi(x, b_i)).$$

S is not an initial segment of n because otherwise some a_j would represent it. But S is obtained as the last step $S = S_m$ in a sequence $S_0, \ldots, S_m$ of subsets $S_k \subseteq n$ where S_0 is an initial segment and for each k there is some $m \in S_k$ such that $m + 1 \notin S_k$ and $S_{k+1} = (S_k \setminus \{m\}) \cup \{m + 1\}$. Since S_0 is represented but S_m is not, there is some k such that S_k is represented but S_{k+1} is not. Let $U = S_k \cap S_{k+1}$, $V = n \setminus (S_k \cup S_{k+1})$ and let $m \in S_k$ be such that $S_k = U \cup \{m\}$ and $S_{k+1} = U \cup \{m+1\}$. If $\psi(x) = \bigwedge_{i \in U} \varphi(x, b_i) \wedge \bigwedge_{i \in V} \neg\varphi(x, b_i)$, it follows that since S_k is represented,

$$\models \exists x(\psi(x) \wedge \varphi(x, b_m) \wedge \neg\varphi(x, b_{m+1}))$$

but since S_{k+1} is not represented,

$$\models \neg\exists x(\psi(x) \wedge \varphi(x, b_{m+1}) \wedge \neg\varphi(x, b_m)).$$

Hence if $\theta(x, y) = \psi(x) \wedge \varphi(x, y)$ we have that

$$\theta(\mathfrak{C}, b_{m+1}) \subsetneq \theta(\mathfrak{C}, b_m).$$

By indiscernibility, for all rational numbers $m \leq q < q' \leq m+1$, $\theta(\mathfrak{C}, b_{q'}) \subsetneq \theta(\mathfrak{C}, b_q)$ which implies that $\theta(x, y)$ has the strict order property. ⊣

DEFINITION 2.19. Let $k \geq 2$ be a natural number. It is said that $\varphi(x, y) \in L$ has the *k-tree property* if there is a tree $(a_s : s \in \omega^{<\omega})$ such that

1. For each $f \in \omega^\omega$, $\{\varphi(x, a_{f \restriction n}) : n < \omega\}$ is consistent.
2. For each $s \in \omega^{<\omega}$ the set $\{\varphi(x, a_{s^\frown i}) : i < \omega\}$ is k-inconsistent, that is, every subset of k elements is inconsistent.

The formula φ has the *tree property* if it has the k-tree property for some k.

PROPOSITION 2.20. *If $\varphi(x, y)$ has the strict order property, then*

$$\psi(x; y_1 y_2) = \neg\varphi(x, y_1) \wedge \varphi(x, y_2)$$

has the 2-tree property.

PROOF. By the strict order property, there is a sequence $(a_p : p < \mathbb{Q})$ such that $\varphi(\mathfrak{C}, a_p) \subsetneq \varphi(\mathfrak{C}, a_q)$ for $p < q \in \mathbb{Q}$. We prove the existence of parameters $b_s = b_s^1 b_s^2$, $(s \in \omega^{<\omega})$ witnessing the 2-tree property of $\psi(x; y_1, y_2)$. The construction is done by induction on the length of s in such a way that for each $s \in \omega^{<\omega}$ there are $p_s < q_s \in \mathbb{Q}$ with $a_{p_s} = b_s^1$ and $a_{q_s} = b_s^2$ and $p_t < p_s < q_s < q_t$ if $t \subsetneq s$. We start with $p_\emptyset = 0$ and $q_\emptyset = 1$. To extend the branch finishing in $s \in \omega^{<\omega}$ it is enough to pick two increasing sequences of rational numbers $(p_{s^\frown i} : i < \omega)$ and $(q_{s^\frown i} : i < \omega)$ such that $p_s < p_{s^\frown i} < q_{s^\frown i} < p_{s^\frown i+1} < q_s$. ⊣

PROPOSITION 2.21. *Any formula with the tree property is unstable.*

PROOF. Assume $\varphi = \varphi(x, y)$ has the k-tree property. Chose λ such that $\lambda^\omega > 2^\omega$ and $\lambda^\omega > \lambda$. By compactness, there is a tree $(a_s : s \in \lambda^{<\omega})$ such that for each $s \in \lambda^{<\omega}$, $\{\varphi(x, a_{s^\frown i}) : i < \lambda\}$ is k-inconsistent and for each $f \in \lambda^\omega$, $\pi_f(x) = \{\varphi(x, a_{f \restriction n}) : n < \omega\}$ is consistent. Choose for each $f \in \lambda^\omega$ a subset $I_f \subseteq \lambda^\omega$ such that $f \in I_f$ and $p_f(x) = \bigcup_{g \in I_f} \pi_g(x)$ is a maximally consistent union of types π_g. By k-inconsistency I_f is a k-branching tree of height ω and hence $|I_f| \leq 2^\omega$. Since $\lambda^\omega > 2^\omega$, $\{p_f(x) : f \in \lambda^\omega\}$ has cardinality λ^ω. Since it is a set of pairwise incompatible φ-types over a set of parameters $\{a_s : s \in \lambda^{<\omega}\}$ of cardinality λ, we conclude that φ is not λ-stable. ⊣

DEFINITION 2.22. The theory T is *λ-stable*, or *stable in λ*, if $|S_n(A)| \leq \lambda$ whenever $|A| \leq \lambda$ and $n < \omega$. It is enough to check this for $n = 1$. T is *stable* if all formulas are stable in T, otherwise it is *unstable*.

PROPOSITION 2.23. *The following are equivalent*:

1. *T is stable.*
2. *T is λ-stable for all λ such that $\lambda^{|T|} = \lambda$.*
3. *T is λ-stable for some λ.*

PROOF. $1 \Rightarrow 2$. Clear since if x is a n-tuple $|S_n(A)| \leq |\prod_{\varphi(x,y)} S_\varphi(A)|$. $3 \Rightarrow 1$. If T is stable in λ, then every formula is λ-stable. ⊣

DEFINITION 2.24. T is *simple* if it does not have formulas with the tree property. It is said that T has *the independence property* if some formula has the independence property in T, otherwise we say T has *nip* or it is *dependent*. Finally, it is said that T has *the strict order property* if some formula has the strict order property in T.

REMARK 2.25. *We have seen that*

1. *T is unstable if and only if T has the independence property or it has the strict order property.*
2. *Simple theories do not have the strict order property.*
3. *Stable theories are simple.*
4. *T is stable if and only if T is simple and does not have the independence property.*

REMARK 2.26. 1. *T is stable if and only if $T(A)$ is stable.*
2. *T is simple if and only if $T(A)$ is simple.*

PROOF. If $\varphi(x, y) \in L$ has the order property in T, it also has the order property in $T(A)$. On the other hand, if $\varphi(x, y, z) \in L$, $a \in A$, and $\varphi(x, y, a)$ has the order property in $T(A)$, then $\varphi(x; yz)$ has the order property in T. This proves 1. A similar argument with the k-tree property can be used for 2. $\dashv$

REMARK 2.27. 1. *T is stable if and only if T^{eq} is stable.*
2. *T is simple if and only if T^{eq} is simple.*

PROOF. It is evident that if T^{eq} is stable or simple then T is also stable or simple respectively. For the opposite direction use Lemma 1.8. $\dashv$

Chapter 3

Δ-TYPES AND THE LOCAL RANK $\mathrm{D}(\pi, \Delta, k)$

We generalize the notions of φ-formula and φ-type to finite sets of formulas Δ.

DEFINITION 3.1. Let $\Delta = \{\varphi_i(x, y_i) : 1 \leq i \leq n\}$ where $\varphi_i(x, y_i) \in L$ for each i. A *Δ-formula* over A is a formula of the form $\varphi_i(x, a)$ or $\neg\varphi_i(x, a)$ with $a \in A$. A *Δ-type over A* is a consistent set of Δ-formulas over A. A Δ-type $p(x)$ over A is complete if for all $i = 1, \ldots, n$ for every $a \in A$, either $\varphi_i(x, a) \in p$ or $\neg\varphi_i(x, a) \in p$. The set of all complete Δ-types over A is $S_\Delta(A)$. We endow $S_\Delta(A)$ with a compact Hausdorff totally disconnected topology. A basis of clopen sets is given by all sets of the form

$$[\psi] = \{p \in S_\Delta(A) : p \vdash \psi\}$$

for any boolean combination $\psi = \psi(x)$ of Δ-formulas over A. All these notions also apply to the case $A = \mathfrak{C}$.

DEFINITION 3.2. Let $\Delta = \{\varphi_i(x, y_i) : 1 \leq i \leq n\}$ and let $k < \omega$. The local D-rank with respect to Δ and k is defined inductively for any set of formulas $\pi = \pi(x)$ by the following clauses:

1. $\mathrm{D}(\pi, \Delta, k) \geq 0$ if and only if π is consistent.
2. $\mathrm{D}(\pi, \Delta, k) \geq \alpha + 1$ if and only if there is some i $(1 \leq i \leq n)$ and there is some sequence $(a_j : j < \omega)$ such that $\{\varphi_i(x, a_j) : j < \omega\}$ is k-inconsistent and for all $j < \omega$, $\mathrm{D}(\pi(x) \cup \{\varphi_i(x, a_j)\}, \Delta, k) \geq \alpha$.
3. $\mathrm{D}(\pi, \Delta, k) \geq \alpha$ if and only if $\mathrm{D}(\pi, \Delta, k) \geq \beta$ for all $\beta < \alpha$ if α is a limit ordinal.

Observe that $\{\alpha : \mathrm{D}(\pi, \Delta, k) \geq \alpha\}$ is an initial segment of the ordinals. If π is inconsistent we set $\mathrm{D}(\pi, \Delta, k) = -1$; otherwise $\mathrm{D}(\pi, \Delta, k)$ is the supremum of all α such that $\mathrm{D}(\pi, \Delta, k) \geq \alpha$. Hence, if $\mathrm{D}(\pi, \Delta, k) \geq \alpha$ for all α then $\mathrm{D}(\pi, \Delta, k) = \infty$. In case $\Delta = \{\varphi(x, y)\}$ we use the notation $\mathrm{D}(\pi, \varphi, k)$.

REMARK 3.3. 1. *Assume* $\pi(x) \vdash \pi'(x)$, $\Delta \subseteq \Delta'$ *and* $k \leq k'$. *Then* $\mathrm{D}(\pi(x), \Delta, k) \leq \mathrm{D}(\pi'(x), \Delta', k')$.

2. *If* $\pi(x)$ *and* $\pi'(x)$ *are equivalent, then* $\mathrm{D}(\pi(x), \Delta, k) = \mathrm{D}(\pi'(x), \Delta, k)$.
3. *For any set* $\pi(x, y)$ *of formulas over* A, *for any* Δ, k, *and* α,

$$\{a : \mathrm{D}(\pi(x, a), \Delta, k) \geq \alpha\}$$

is type-definable over A.

PROOF. 2 follows from 1, and to prove 1 one shows by induction on α that

$$D(\pi(x), \Delta, k) \geq \alpha \Rightarrow D(\pi'(x), \Delta', k') \geq \alpha.$$

3 is easily proved by induction on α. ⊣

LEMMA 3.4. *Let $\Delta = \{\varphi_1(x, y_1), \dots, \varphi_n(x, y_n)\}$ where $\varphi_i(x, y_i) \in L$ for every i. There is a formula*

$$\psi_\Delta = \psi_\Delta(x; y_1, \dots, y_n, z, z_1, \dots, z_{2n}) \in L$$

such that

1. *For each A with $|A| \geq 2$ for each Δ-formula $\varphi(x)$ over A there is a tuple $a \in A$ such that $\varphi(x) \equiv \psi_\Delta(x; a)$.*
2. *For each A for each tuple $a \in A$ such that $\psi_\Delta(x; a)$ is consistent, there is a Δ-formula $\varphi(x)$ over A such that $\varphi(x) \equiv \psi_\Delta(x; a)$.*

PROOF. Take as $\psi_\Delta(x; y_1, \dots, y_n, z, z_1, \dots, z_{2n})$ the following formula:

$$\bigwedge_{i=1}^{n}(z = z_i \rightarrow \varphi_i(x, y_i)) \wedge \bigwedge_{i=1}^{n}(z = z_{n+i} \rightarrow \neg\varphi_i(x, y_i)) \wedge$$
$$\bigvee_{i=1}^{2n}(z = z_i) \wedge \bigwedge_{1 \leq i < j \leq 2n} \neg(z = z_i \wedge z = z_j).$$

Choose $a_0, a_1 \in A$ such $a_0 \neq a_1$. Then for each $a \in A$, $\varphi_i(x, a)$ is equivalent to

$$\psi_\Delta(x; b_1, \dots, b_n, c, c_1, \dots, c_{2n})$$

where $b_i = a$ for all $i = 1, \dots, n$, $c = a_0 = c_i$ and $c_j = a_1$ for $j \neq i$; and $\neg\varphi_i(x, a)$ is equivalent to

$$\psi_\Delta(x; b_1, \dots, b_n, c, c_1, \dots, c_{2n})$$

where $b_i = a$ for all $i = 1, \dots, n$, $c = a_0 = c_{n+i}$ and $c_j = a_1$ for $j \neq n + i$ ⊣

REMARK 3.5. *If $\Delta = \{\varphi_1(x, y_1), \dots, \varphi_n(x, y_n)\}$, each $\varphi_i(x, y_i)$ is stable, and ψ_Δ is chosen as in the previous lemma, then ψ_Δ is stable.*

PROOF. By Corollary 2.12. ⊣

COROLLARY 3.6. *For each set $\Delta = \{\varphi_1(x, y_1), \dots, \varphi_n(x, y_n)\}$ of formulas $\varphi_i(x, y_i) \in L$, there is a formula $\psi_\Delta(x, z) \in L$ such that for all $\pi(x)$, for all k,*

$$D(\pi, \Delta, k) = D(\pi, \psi_\Delta, k).$$

PROOF. The formula ψ_Δ is chosen accordingly to Lemma 3.4. By induction on α we see that for each π and k, $D(\pi, \Delta, k) \geq \alpha$ if and only if $D(\pi, \psi_\Delta, k) \geq \alpha$. This is clear for $\alpha = 0$ and follows from the induction hypothesis for limit α. The case $\alpha + 1$ is easy and only requires noticing that Δ is finite and therefore any infinite sequence of Δ-formulas contains an infinite subsequence of instances of a single formula. ⊣

Due to this last result, we will concentrate on the study of $D(\pi, \varphi, k)$ rank. The generalization of the statements to $D(\pi, \Delta, k)$ rank is straightforward.

DEFINITION 3.7. Let $\varphi(x, y) \in L$ and let $k < \omega$. The formula $\varphi(x, a)$ *k-divides over A* if there is a sequence $(a_i : i < \omega)$ such that $\{\varphi(x, a_i) : i < \omega\}$ is k-inconsistent and $a_i \equiv_A a$ for all $i < \omega$. We say that $\varphi(x, a)$ *divides over A* if it k-divides over A for some k. If A is omitted we understand that $A = \emptyset$. Sometimes we say $\varphi(x, a)$ *divides over A with respect to k* instead of saying that $\varphi(x, a)$ k-divides over A.

PROPOSITION 3.8. *Let $(\varphi_i(x, y_i) : i < \alpha)$ be a sequence of L-formulas and let $(k_i : i < \alpha)$ be a sequence of natural numbers. For any partial type $\pi(x)$ over A, the following are equivalent:*

1. *There is a sequence $(b_i : i < \alpha)$ such that $\pi(x) \cup \{\varphi_i(x, b_i) : i < \alpha\}$ is consistent and for each $i < \alpha$, $\varphi_i(x, b_i)$ k_i-divides over $Ab_{<i}$.*
2. *There is a tree $(a_s : s \in \omega^{\leq\alpha})$ such that for each $f \in \omega^\alpha$,*

$$\pi(x) \cup \{\varphi_i(x, a_{f\restriction i+1}) : i < \alpha\}$$

is consistent and for each $i < \alpha$, for each $s \in \omega^i$, $\{\varphi_i(x, a_{s\frown j}) : j < \omega\}$ is k_i-inconsistent.

PROOF. We first prove that 1 implies 2. Observe that a_s plays no role if the length of s is 0 or a limit ordinal. We construct a_s for $s \in \omega^i$ by induction in $i \leq \alpha$ with the additional property that $(a_{s\restriction j+1} : j < i) \equiv_A (b_j : j < i)$. Assume $s \in \omega^i$ and that a_s has already been obtained. Choose c such that

$$(a_{s\restriction j+1} : j < i)c \equiv_A (b_j : j < i)b_i.$$

Then $\varphi_i(x, c)$ k_i-divides over $A' = A\{a_{s\restriction j+1} : j < i\}$ and therefore we can find $a_{s\frown l} \equiv_{A'} c$ for all $l < \omega$ such that $\pi(x) \cup \{\varphi_j(x, a_{s\restriction j+1}) : j < i\} \cup \{\varphi_i(x, a_{s\frown l})\}$ is consistent and $\{\varphi_i(x, a_{s\frown l}) : l < \omega\}$ is k_i-inconsistent.

For the other direction choose first $\lambda > 2^{|T|+|A|+|\alpha|}$. By compactness there are a_s, $(s \in \lambda^{\leq\alpha})$ such that for each $f \in \lambda^\alpha$, $\pi(x) \cup \{\varphi_i(x, a_{f\restriction i+1}) : i < \alpha\}$ is consistent and for each $i < \alpha$, for each $s \in \lambda^i$, $\{\varphi_i(x, a_{s\frown l}) : l < \lambda\}$ is k_i-inconsistent. Observe that by choice of λ, for any $i < \alpha$ for any $s \in \lambda^i$ at least λ of the $a_{s\frown l}$ have the same type over $A\{a_{s\restriction j+1} : j < i\}$. Hence we can prune the tree obtaining a subtree where this happens for all $a_{s\frown l}$. Finally choose a branch $f \in \lambda^\alpha$ and put $b_i = a_{f\restriction i+1}$ for all $i < \alpha$. ⊣

REMARK 3.9. *Assume $\pi(x)$ is a partial type over A, $\varphi(x, y) \in L$, and $k < \omega$. If for each $n < \omega$ there is a sequence $(a_i : i < n)$ such that $\pi(x) \cup \{\varphi(x, a_i) : i < n\}$ is consistent and for each $i < n$, $\varphi(x, a_i)$ k-divides over $Aa_{<i}$, then for each ordinal α there is a sequence $(a_i : i < \alpha)$ such that $\pi(x) \cup \{\varphi(x, a_i) : i < \alpha\}$ is consistent and for each $i < \alpha$, $\varphi(x, a_i)$ k-divides over $Aa_{<i}$.*

PROOF. By compactness and Proposition 3.8. ⊣

LEMMA 3.10. *Let $\pi(x)$ be a partial type over A. $D(\pi(x), \Delta, k) \geq \alpha + 1$ if and only if for some $\varphi(x, y) \in \Delta$, for some a, $D(\pi(x) \cup \{\varphi(x, a)\}, \Delta, k) \geq \alpha$ and $\varphi(x, a)$ k-divides over A.*

PROOF. The direction from right to left is obvious from the definitions of D-rank and dividing. For the other direction, assume $D(\pi(x), \Delta, k) \geq \alpha + 1$. Let $\lambda > 2^{|T|+|A|}$. From point 3 in Remark 3.3 and compactness, we see that there are $\varphi(x, y) \in \Delta$ and $(a_i : i < \lambda)$ such that for each $i < \lambda$, $D(\pi(x) \cup \{\varphi(x, a_i)\}, \Delta, k) \geq \alpha$ and $\{\varphi(x, a_i) : i < \lambda\}$ is k-inconsistent. By choice of λ, there is an infinite subset $I \subseteq \lambda$ such that $a_i \equiv_A a_j$ for all $i, j \in I$. Then it suffices to take $a = a_i$ for some $i \in I$. ⊣

PROPOSITION 3.11. *For any partial type $\pi(x)$ over A, any $\varphi = \varphi(x, y) \in L$, any $k < \omega$, and any ordinal $\alpha \leq \omega$ the following are equivalent:*

1. *$D(\pi(x), \varphi, k) \geq \alpha$.*
2. *There is a sequence $(a_i : i < \alpha)$ such that $\pi(x) \cup \{\varphi(x, a_i) : i < \alpha\}$ is consistent and for each $i < \alpha$, $\varphi(x, a_i)$ k-divides over $Aa_{<i}$.*

PROOF. By induction on α. The case $\alpha = 0$ is obvious. Let us consider the case $\alpha = n + 1$. Assume there is a sequence $(a_i : i < n + 1)$ such that $\pi(x) \cup \{\varphi(x, a_i) : i < n + 1\}$ is consistent and for each $i < n + 1$, $\varphi(x, a_i)$ k-divides over $Aa_{<i}$. By the induction hypothesis $D(\pi(x) \cup \{\varphi(x, a_0)\}, \varphi, k) \geq n$ and by Lemma 3.10 we see that $D(\pi(x), \varphi, k) \geq n+1$. For the other direction, assume now $D(\pi(x), \varphi, k) \geq n + 1$. Again by Lemma 3.10 there is some a_0 such that $\varphi(x, a_0)$ k-divides over A and $D(\pi(x) \cup \{\varphi(x, a_0)\}, \varphi, k) \geq n$. By the induction hypothesis there is a sequence $(b_i : i < n)$ such that $\pi(x) \cup \{\varphi(x, a_0)\} \cup \{\varphi(x, b_i) : i < n\}$ is consistent and for each $i < n$, $\varphi(x, b_i)$ k-divides over $Aa_0b_{<i}$. The case $\alpha = \omega$ follows by the induction hypothesis and Remark 3.9. ⊣

PROPOSITION 3.12. *Fix φ and k.*

1. *If $D(\pi, \varphi, k) \geq \omega$, then $D(\pi, \varphi, k) = \infty$.*
2. *There is a conjunction $\psi(x)$ of formulas from π such that $D(\pi, \varphi, k) = D(\psi, \varphi, k)$.*
3. *$D(\pi(x) \cup \{\psi_1(x) \vee \cdots \vee \psi_n(x)\}, \varphi, k) = \max_{i=1}^{n} D(\pi(x) \cup \{\psi_i(x)\}, \varphi, k)$.*
4. *Every partial type $\pi(x)$ over A can be extended to a complete type $p(x) \in S(A)$ such that $D(\pi, \varphi, k) = D(p, \varphi, k)$.*

PROOF. 1 follows from Proposition 3.11 and Remark 3.9. Hence in 2 it is enough to find this conjunction ψ with $D(\psi, \varphi, k) \not\geq n + 1$ where $n = D(\pi, \varphi, k)$; this can be done by compactness and propositions 3.11 and 3.8.

For 3 we use Proposition 3.11. Assume $\pi(x)$ is a partial type over A and let $\psi_i(x) = \psi_i(x, b_i)$ where $\psi_i(x, y_i) \in L$. Assume $D(\pi(x) \cup \{\psi_1(x) \vee \cdots \vee \psi_n(x)\}, \varphi, k) \geq \alpha$. There is a sequence $(a_l : l < \alpha)$ such that $\pi(x) \cup \{\psi_1(x) \vee \cdots \vee \psi_n(x)\} \cup \{\varphi_l(x, a_l) : l < \alpha\}$ is consistent and for each $l < \alpha$, $\varphi(x, a_l)$ k-divides over $Ab_1, \ldots, b_n, a_{<l}$. Clearly, for some i, $\pi(x) \cup \{\psi_i(x)\} \cup \{\varphi_l(x, a_l) :$

$l < \alpha\}$ is consistent. Since $\varphi(x, a_l)$ also k-divides over $Ab_i a_{<l}$ we conclude that $D(\pi(x) \cup \{\psi_i(x)\}, \varphi, k) \geq \alpha$.

For 4, use 3 to guarantee the consistency of

$$\pi(x) \cup \{\neg\psi(x) : \psi(x) \in L(A) \text{ and } D(\pi(x) \cup \{\psi(x)\}, \varphi, k) < D(\pi(x), \varphi, k)\}$$

and take as $p(x)$ any complete type over A extending this consistent set of formulas. ⊣

PROPOSITION 3.13. 1. *$\varphi(x, y)$ has the k-tree property if and only if* $D(x = x, \varphi, k) = \infty$.

2. *T is simple if and only if* $D(x = x, \varphi, k) < \omega$ *for all φ and k.*

PROOF. The first point follows from propositions 3.11 and 3.8 and the second one follows directly from the first. ⊣

REMARK 3.14. *For all $\varphi(x, y) \in L$, $n, k < \omega$, for all A,*

$$\{(a, b) : D(\text{tp}(a/Ab), \varphi, k) \geq n\}$$

is type-definable over A.

PROOF. Let us write $p_{ab}(x) = \text{tp}(a/b)$. By propositions 3.11 and 3.8, $D(p_{ab}(x), \varphi, k) \geq n$ if and only if there is a tree $(a_s : s \in \omega^{\leq n})$ such that for each $f \in \omega^n$, $p_{ab}(x) \cup \{\varphi(x, a_{f\restriction i+1}) : i < n\}$ is consistent and for each $i < n$, for each $s \in \omega^i$, $\{\varphi(x, a_{s^\frown j}) : j < \omega\}$ is k-inconsistent. Consistency of $p_{ab}(x) \cup \{\varphi(x, a_{f\restriction i+1}) : i < n\}$ can be expressed by:

$$\exists x(\bigwedge_{i<n} \varphi(x, a_{f\restriction i+1}) \wedge \bigwedge_{\psi(x,y)\in L(A)} (\psi(x, b) \leftrightarrow \psi(a, b))).$$

⊣

COROLLARY 3.15. *If T is simple, every partial type $\pi(x)$ over A has an extension $p(x) \in S(A)$ with maximal local* D*-rank, that is, if $q(x) \in S(A)$ is an extension of π such that $D(q(x), \varphi, k) \geq D(p(x), \varphi, k)$ for all formulas $\varphi = \varphi(x, y) \in L$, for all $k < \omega$, then $D(q(x), \varphi, k) = D(p(x), \varphi, k)$ for all φ, k.*

PROOF. It is an application of Zorn's Lemma. To obtain an upper bound for an ascending chain, use Remark 3.14 and compactness. ⊣

Chapter 4

FORKING

DEFINITION 4.1. Let $\pi(x)$ be a set of formulas over B. We say that $\pi(x)$ *divides over* A if π implies a formula $\varphi(x, a)$ which divides over A. Notice that we may always assume that $a \in B$ and that $\varphi(x, a)$ is a conjunction of formulas in $\pi(x)$.

REMARK 4.2. 1. *$\varphi(x, a)$ divides over A if and only if the set $\{\varphi(x, a)\}$ divides over A.*
2. *If $\pi(x)$ is inconsistent, it divides over A.*
3. *A partial type $\pi(x)$ over* $\mathrm{acl}(A)$ *does not divide over A.*
4. *$\pi(x, a)$ divides over A if and only if for some A-indiscernible sequence $(a_i : i < \omega)$ with $a_0 = a$, the set of formulas $\bigcup_{i<\omega} \pi(x, a_i)$ is inconsistent.*
5. $\mathrm{acl}(A) = \{a : \mathrm{tp}(a/Aa)$ *does not divide over* $A\}$.
6. *$\pi(x)$ divides over A if and only if it divides over* $\mathrm{acl}(A)$.
7. *$\pi(x)$ divides over A if and only if it divides over some model $M \supseteq A$.*

PROOF. For 2 take as $\varphi(x, y)$ the formula $x \neq x$. For 4 use Ramsey's Theorem to obtain indiscernibility. For 5 consider the formula $x = a$. For 6 and 7 use Corollary 1.7. ⊣

DEFINITION 4.3. The set of formulas $\pi(x)$ *forks over* A if for some n there are formulas $\varphi_1(x, a_1), \ldots, \varphi_n(x, a_n)$ such that $\pi(x) \vdash \varphi_1(x, a_1) \vee \cdots \vee \varphi_n(x, a_n)$ and every $\varphi_i(x, a_i)$ divides over A. The formula $\varphi(x, a)$ forks over A if the set $\{\varphi(x, a)\}$ forks over A.

A *nonforking extension* of a type $p(x) \in S(A)$ over $B \supseteq A$ is a type $q(x) \in S(B)$ that extends p and does not fork over A. If an extension q of p forks over A it is called a *forking extension* of p.

REMARK 4.4. 1. *If $\pi(x)$ divides over A, then it forks over A.*
2. *If $\pi(x)$ is finitely satisfiable in A, then it does not fork over A.*
3. *$\pi(x)$ forks over A if and only if a conjunction of formulas in $\pi(x)$ forks over A.*
4. *If a partial type $\pi(x)$ over B does not fork over A, then it can be extended to a complete type over B which does not fork over A.*
5. *$\pi(x)$ forks over A if and only if it forks over* $\mathrm{acl}(A)$.

6. *Let $\pi(x)$ be a partial type over A. If $\pi(x)$ is algebraic, then no extension of $\pi(x)$ forks over A.*

PROOF. The first three points follow directly from the definitions. For 4 check the consistency of $\pi(x) \cup \{\neg\varphi(x) : \varphi(x) \in L(B) \text{ forks over } A\}$ and take as p any complete type over B extending this partial type. For 5 use point 6 of Remark 4.2.

6. Assume $\pi'(x)$ extends $\pi(x)$ and forks over A. By 5, $\pi'(x)$ forks over $\operatorname{acl}(A)$. But this contradicts 2 since all realizations of $\pi'(x)$ lie in $\operatorname{acl}(A)$. ⊣

EXAMPLE 4.5. A formula that forks but does not divide. It is taken from [15], Example 2.11. In a circle consider the ternary relation: $B(x, y, z)$ if and only if y lies between x and z (clockwise) and the arc from x to z is shorter than the arc from z to x. If we choose a_0, a_1, a_2 dividing the circle into three equal parts and $B(a_0, a_1, a_2)$, then the disjunction $B(a_0, x, a_1) \vee B(a_1, x, a_2) \vee B(a_2, x, a_0)$ defines the whole circle and hence it does not divide over $\emptyset$. But each $B(a_i, x, a_j)$ divides over $\emptyset$. Therefore $x = x$ forks but does not divide over $\emptyset$.

PROPOSITION 4.6. *Let $A \subseteq M$ and assume for each finite $B \subseteq M$ every finitary type over AB is realized in M. If $p(x) \in S(M)$, then p divides over A if and only if p forks over A.*

PROOF. Assume $p(x)$ forks over A. For some $\theta(x, y) \in L$, for some finite tuple $a \in M$ there are $\varphi_1(x, y_1), \dots, \varphi_n(x, y_n) \in L$ and finite tuples $b_1, \dots, b_n$ such that $p(x) \vdash \theta(x, a)$, each $\varphi_i(x, b_i)$ divides over A, and

$$\theta(x, a) \vdash \varphi_1(x, b_1) \vee \dots \vee \varphi_n(x, b_n).$$

By the assumption on M we may find tuples $a_1, \dots, a_n \in M$ such that $a_1, \dots, a_n \equiv_{Aa} b_1, \dots, b_n$. Then $\theta(x, a) \vdash \varphi_1(x, a_1) \vee \dots \vee \varphi_n(x, a_n)$ and each $\varphi_i(x, a_i)$ divides over A. Since $p(x) \vdash \varphi_i(x, a_i)$ for some i, we conclude that $p(x)$ divides over A. ⊣

LEMMA 4.7. *The following are equivalent:*

1. $\operatorname{tp}(a/Ab)$ *does not divide over A.*
2. *For every infinite A-indiscernible sequence I such that $b \in I$, there is some $a' \equiv_{Ab} a$ such that I is Aa'-indiscernible.*
3. *For every infinite A-indiscernible sequence I such that $b \in I$, there is some $J \equiv_{Ab} I$ such that J is Aa-indiscernible.*

PROOF. The equivalence of 2 and 3 follows by conjugation. It is clear that 3 implies 1. We prove that 1 implies 2. We may assume that A is empty, that $I = (b_i : i < \omega)$ and that $b = b_0$. Let $p(x, b) = \operatorname{tp}(a/b)$ and let $\Gamma(x, (x_i : i < \omega))$ be the set of all formulas

$$\varphi(x, x_0, \dots, x_n) \leftrightarrow \varphi(x, x_{i_0}, \dots, x_{i_n})$$

for all $\varphi \in L$ and $i_0 < \cdots < i_n < \omega$. This set expresses the fact that $(x_i : i < \omega)$ is x-indiscernible. It will be enough to prove that $p(x, b) \cup \Gamma(x, (b_i : i < \omega))$ is consistent. By 1, $q(x) = \bigcup_{i<\omega} p(x, b_i)$ is consistent. Let $c \models q$ and let Γ_0 a finite subset of Γ. By Ramsey's Theorem, there is an order preserving $f : \omega \to \omega$ such that $\models \Gamma_0(c, (b_{f(i)} : i < \omega))$. By indiscernibility $(b_i : i < \omega) \equiv (b_{f(i)} : i < \omega)$ and therefore we can find c' such that $c'(b_i : i < \omega) \equiv c(b_{f(i)} : i < \omega)$. Clearly $c' \models q(x) \cup \Gamma_0(x, (b_i : i < \omega))$. ⊣

PROPOSITION 4.8. *If* $\mathrm{tp}(a/B)$ *does not divide over* $A \subseteq B$ *and* $\mathrm{tp}(b/Ba)$ *does not divide over* Aa, *then* $\mathrm{tp}(ab/B)$ *does not divide over* A.

PROOF. It is an easy application of point 3 of Lemma 4.7. ⊣

PROPOSITION 4.9. *If* $\varphi(x, a)$ k*-divides over* A *and* $\mathrm{tp}(b/Aa)$ *does not divide over* A, *then* $\varphi(x, a)$ k*-divides over* Ab.

PROOF. Let $I = (a_i : i < \omega)$ be an A-indiscernible sequence such that $a = a_0$ and $\{\varphi(x, a_i) : i < \omega\}$ is k-inconsistent. By Lemma 4.7 there is some $J \equiv_{Aa} I$ which is Ab-indiscernible. Then J witnesses that $\varphi(x, a)$ divides over Ab with respect to k. ⊣

DEFINITION 4.10. A *dividing chain* for $\varphi(x, y)$ is a sequence $(a_i : i < \alpha)$ such that $\{\varphi(x, a_i) : i < \alpha\}$ is consistent and for every $i < \alpha$, $\varphi(x, a_i)$ divides over $a_{<i}$. If $\varphi(x, a_i)$ k_i-divides over $a_{<i}$, we say that it is a dividing chain with respect to $(k_i : i < \alpha)$. We say that $\varphi(x, y)$ *divides* α *times* (with respect to $(k_i : i < \alpha)$) if there is a dividing chain of length α for $\varphi(x, y)$ (with respect to $(k_i : i < \alpha)$).

REMARK 4.11.
1. $\varphi(x, y)$ *divides* ω *times with respect to* k *if and only if it has the tree property with respect to* k.
2. *If* $\varphi(x, y)$ *divides* n *times with respect to* k *for every* $n < \omega$, *then it divides* α *times with respect to* k *for every ordinal* α.
3. *If* $\varphi(x, y)$ *divides* ω_1 *times, then for some* $k < \omega$, $\varphi(x, y)$ *divides* ω *times with respect to* k.

PROOF. For 1 and 2 use Remark 3.9 and Proposition 3.8. For 3 note that if $\varphi(x, a_i)$ k_i-divides over $a_{<i}$ for all $i < \omega_1$, then some k_i is being used infinitely many times. ⊣

REMARK 4.12. *Clearly, simplicity is equivalent to the nonexistence of formulas which divide* ω *times with respect to some fixed* k *and also to the non existence of formulas which divide* ω_1 *times (with respect to possibly varying* k*).*

PROPOSITION 4.13. *The following conditions are equivalent to the simplicity of* T. *Here all types are assumed to be finitary.*

1. *For every type* $p(x) \in S(A)$ *there is a* $B \subseteq A$ *such that* $|B| \leq |T|$ *and* p *does not divide over* B.
2. *There is some cardinal* κ *such that for every type* $p(x) \in S(A)$ *there is a* $B \subseteq A$ *such that* $|B| \leq \kappa$ *and* p *does not divide over* B.

3. *There is no increasing chain $(p_i(x) : i < |T|^+)$ of types $p_i(x) \in S(A_i)$ such that for every $i < |T|^+$, p_{i+1} divides over A_i.*
4. *For some cardinal $\kappa \geq |T|$ there is no increasing chain $(p_i(x) : i < \kappa^+)$ of types $p_i(x) \in S(A_i)$ such that for every $i < \kappa^+$, p_{i+1} divides over A_i.*

PROOF. Simplicity implies 1, since if $p \in S(A)$ divides over every subset of A of cardinality $\leq |T|$, then we can inductively construct a sequence of formulas $(\varphi_i(x, y_i) : i < |T|^+)$ and a sequence $(a_i : i < |T|^+)$ of parameters $a_i \in A$ such that $\varphi_i(x, a_i) \in p$ and $\varphi_i(x, a_i)$ divides over $a_{<i}$. Clearly one formula $\varphi_i(x, y_i)$ appears ω_1 times in the sequence and this contradicts simplicity. It is clear that 1 implies 2 and that 3 implies 4. To show $1 \Rightarrow 3$, observe that if the increasing chain $(p_i(x) : i < |T|^+)$ is given and we set $A = \bigcup_i A_i$ and $p = \bigcup_i p_i$, then $p(x) \in S(A)$ divides over every subset of A of cardinality $\leq |T|$. The same argument proves $2 \Rightarrow 4$. It remains only to prove simplicity assuming 4. If T is not simple, then some formula $\varphi(x, y)$ divides κ^+ times. Let $(a_i : i < \kappa^+)$ be a witness of this. Let a be a realization of $\{\varphi(x, a_i) : i < \kappa^+\}$, let $A_i = \{a_j : j < i\}$ and let $p_i = \mathrm{tp}(a/A_i)$. The chain $(p_i : i < \kappa^+)$ contradicts point 4. ⊣

LEMMA 4.14. *Let $\Delta = \{\varphi_1(x, y_1), \ldots, \varphi_n(x, y_n)\}$. Assume $\mathrm{D}(\pi(x) \restriction A, \Delta, k) < \omega$ and $\pi(x) \vdash \varphi_1(x, a_1) \vee \cdots \vee \varphi_n(x, a_n)$ where every $\varphi_i(x, a_i)$ divides over A with respect to k. Then $\mathrm{D}(\pi(x), \Delta, k) < \mathrm{D}(\pi(x) \restriction A, \Delta, k)$.*

PROOF. By Proposition 3.12 there is some i such that

$$\mathrm{D}(\pi(x), \Delta, k) \leq \mathrm{D}(\pi(x) \restriction A \cup \{\varphi_i(x, a_i)\}, \Delta, k).$$

Let $m = \mathrm{D}(\pi(x) \restriction A \cup \{\varphi_i(x, a_i)\}, \Delta, k)$. By Lemma 3.10 $\mathrm{D}(\pi(x) \restriction A, \Delta, k) \geq m + 1$. ⊣

PROPOSITION 4.15. *Simplicity is also equivalent to the conditions in Proposition* 4.13 *if we replace dividing by forking.*

PROOF. Point 4 from Proposition 4.13 stated for forking (instead of dividing) implies its original version. The arguments in the proof of Proposition 4.13 showing that 1 implies 2 and 3 and that any of 2 and 3 implies 4 adapt to its version with forking. Moreover it is pretty clear that 3 implies 1 in any version. Hence it will be enough to prove that simple theories verify point 3 in this new version for forking. Assume $(p_i(x) : i < |T|^+)$ is an increasing chain of types $p_i(x) \in S(A_i)$ such that p_{i+1} forks over A_i for all $i < |T|^+$. This means that for all $i < |T|^+$ we can find some $\varphi_1^i(x), \ldots, \varphi_{n_i}^i(x)$ and some numbers $k_{1i}, \ldots, k_{n_i i}$ such that $p_{i+1}(x) \vdash \varphi_1^i(x) \vee \cdots \vee \varphi_{n_i}^i(x)$ and each $\varphi_j^i(x)$ k_{ji}-divides over A_i. Taking the maximum of the numbers k_{ji} for any fixed i, we may assume that they are all equal to some k_i. By counting types, numbers and formulas, we may assume that there are $n, k < \omega$ and some $\varphi_1(x, y_1), \ldots, \varphi_n(x, y_n) \in L$ such that for all $i < |T|^+$, $n = n_i$, $k = k_i$ and there are tuples $a_1^i, \ldots, a_n^i \in A_{i+1}$ for which $\varphi_i(x, a_j^i) = \varphi_j^i(x)$. Let $\Delta = \{\varphi_1(x, y_1), \ldots, \varphi_n(x, y_n)\}$. By Lemma 4.14

$D(p_i(x), \Delta, k) > D(p_{i+1}(x), \Delta, k)$ for all $i < |T|^+$, which is a contradiction since the local D-rank is finite in a simple theory. ⊣

COROLLARY 4.16. *If T is simple and $p(x) \in S(A)$, then p does not fork over A. Hence, for any $B \supseteq A$ there is a nonforking extension $q(x) \in S(B)$ of p.*

PROOF. It is enough to check the first assertion for finitary types and this case follows from Proposition 4.15. The rest is point 4 in Remark 4.4. ⊣

Chapter 5

INDEPENDENCE

Definition 5.1. We say that A is *independent of B over C* (written $A \mathop{\smile\!\!\!\!\!\!|}_C B$) if for every finite tuple $a \in A$, $\mathrm{tp}(a/BC)$ does not fork over C. $A \not\mathop{\smile\!\!\!\!\!\!|}_C B$ means that not $A \mathop{\smile\!\!\!\!\!\!|}_C B$. In the case $C = \emptyset$ we write $A \mathop{\smile\!\!\!\!\!\!|} B$ and $A \not\mathop{\smile\!\!\!\!\!\!|} B$.

Remark 5.2. *If instead of sets A, B, C we put partially, or everywhere, tuples or sequences a, b, c in the independence relation we mean the independence of the enumerated sets. But it is easy to prove that $A \mathop{\smile\!\!\!\!\!\!|}_C B$ if and only if* $\mathrm{tp}(a/BC)$ *does not fork over C for some (every) enumeration a of A.*

Remark 5.3. *The independence relation always has the following properties:*

1. *Invariance: If $f \in \mathrm{Aut}(\mathfrak{C})$ and $A \mathop{\smile\!\!\!\!\!\!|}_C B$, then $f(A) \mathop{\smile\!\!\!\!\!\!|}_{f(C)} f(B)$.*
2. *Normality: $A \mathop{\smile\!\!\!\!\!\!|}_C B$ if and only if $A \mathop{\smile\!\!\!\!\!\!|}_C CB$ if and only if $AC \mathop{\smile\!\!\!\!\!\!|}_C B$.*
3. *Finite character: If $a \mathop{\smile\!\!\!\!\!\!|}_C b$ for all finite tuples $a \in A$, $b \in B$, then $A \mathop{\smile\!\!\!\!\!\!|}_C B$.*
4. *Base monotonicity: If $A \mathop{\smile\!\!\!\!\!\!|}_C B$ and $B' \subseteq B$, then $A \mathop{\smile\!\!\!\!\!\!|}_{CB'} B$.*
5. *Monotonicity: If $A \mathop{\smile\!\!\!\!\!\!|}_C B$, $A' \subseteq A$ and $B' \subseteq B$, then $A' \mathop{\smile\!\!\!\!\!\!|}_C B'$.*
6. *Anti-reflexivity: If $A \mathop{\smile\!\!\!\!\!\!|}_B A$, then $A \subseteq \mathrm{acl}(B)$.*
7. *Algebraic closure: $\mathrm{acl}(A) \mathop{\smile\!\!\!\!\!\!|}_A B$.*

Proof. Only the implication $A \mathop{\smile\!\!\!\!\!\!|}_C B \Rightarrow AC \mathop{\smile\!\!\!\!\!\!|}_C B$ needs some checking. For this note that if $\varphi(x, y, z) \in L$, $c \in C$, and $\varphi(x, y, d)$ divides over C then also $\varphi(x, c, d)$ divides over C. For point 6 use point 5 of Remark 4.2. Finally, for point 7 use point 6 of Remark 4.4. ⊣

Proposition 5.4. *Let $A \subseteq M$ and assume for every finite $B \subseteq M$, every finitary type over AB is realized in M. If $a \mathop{\smile\!\!\!\!\!\!|}_A M$ and $b \mathop{\smile\!\!\!\!\!\!|}_{Aa} M$, then $ab \mathop{\smile\!\!\!\!\!\!|}_A M$.*

Proof. By Lemma 4.8 and Proposition 4.6. ⊣

Proposition 5.5 (Local character). *Let T be simple. For any B, C there is some $A \subseteq B$ such that $|A| \leq |T| + |C|$ and $C \mathop{\smile\!\!\!\!\!\!|}_A B$.*

Proof. This is clear for finite C by Proposition 4.15. For the general case, choose first some $A_X \subseteq B$ such that $|A_X| \leq |T|$ and $X \mathop{\smile\!\!\!\!\!\!|}_{A_X} B$ for each finite

$X \subseteq C$ and let A be the union of all these sets A_X. Then $|A| \leq |T| + |C|$ and $C \mathop{\smile\!\!\!\!\!\!|}_A B$. ⊣

PROPOSITION 5.6 (Closedness). *The set of all complete types $p(x) \in S(B)$ which do not fork over A is a closed set in $S(B)$.*

PROOF. Let $\pi(x)$ be the set of all negations $\neg\varphi(x)$ of all formulas $\varphi(x) \in L(B)$ which fork over A. Then $p(x) \in S(B)$ does not fork over A if and only if p extends π. ⊣

PROPOSITION 5.7 (Extension). *Let T be simple and let a be a possibly infinite tuple. For any B, there is some $a' \equiv_A a$ such that $a' \mathop{\smile\!\!\!\!\!\!|}_A B$.*

PROOF. By Corollary 4.16. The infinite case is also covered. ⊣

REMARK 5.8. *A type $p(x) \in S(B)$ which does not fork over $A \subseteq B$ has also a global nonforking extension $\mathfrak{p}(x) \in S(\mathfrak{C})$ which does not fork over A. Therefore, in a simple theory any type has a global nonforking extension.*

PROOF. The same argument as for a nonforking extension over a small set. ⊣

DEFINITION 5.9. Let $(I, <)$ be a linearly ordered set. The sequence $(a_i : i \in I)$ is *A-independent* (or *independent over A*) if for every $i \in I$,

$$a_i \mathop{\smile\!\!\!\!\!\!|}_A a_{<i}.$$

A *Morley sequence over A* is a sequence $(a_i : i \in I)$ which is A-independent and A-indiscernible. It is said to be a Morley sequence in the type $p \in S(A)$ if it is a Morley sequence over A and every a_i realizes p.

REMARK 5.10. *Let $(I, <)$ be an infinite linearly ordered set and let $(a_i : i \in I)$ be an A-indiscernible sequence. The sequence is nontrivial (i.e., $a_i \neq a_j$ for all $i \neq j$) if and only if $\operatorname{tp}(a_i/A)$ is nonalgebraic.*

LEMMA 5.11. *If $p(x) \in S(B)$ does not fork over $A \subseteq B$, there is a Morley sequence $(a_i : i < \omega)$ in p which is moreover a Morley sequence over A.*

PROOF. Let α be the length of x, let $\kappa = |B| + |T| + |\alpha|$ and $\lambda = \beth_{(2^\kappa)^+}$. Since $p(x)$ does not fork over A, one can construct a sequence $(a_i : i < \lambda)$ of realizations a_i of p such that $a_i \mathop{\smile\!\!\!\!\!\!|}_A Ba_{<i}$. For this we choose a global extension $\mathfrak{p}$ of p which does not fork over A (see Remark 5.8) and we take as a_i a realization of $\mathfrak{p} \restriction Ba_{<i}$. By Proposition 1.6 there is a B-indiscernible sequence $(b_i : i < \omega)$ of realizations of p such that for each $n < \omega$ there are $i_0 < \cdots < i_n < \lambda$ such that

$$b_0, \ldots, b_n \equiv_B a_{i_0}, \ldots, a_{i_n}.$$

Since $(a_i : i < \lambda)$ is A-independent, it follows that $(b_i : i < \omega)$ is also A-independent and hence it is a Morley sequence over A. But $(a_i : i < \lambda)$ is B-independent too and this also transfers to $(b_i : i < \omega)$. Hence it is a Morley sequence in p. ⊣

REMARK 5.12. *Let $p(x) \in S(A)$. If there is a Morley sequence $(a_i : i < \omega)$ in p, then for any linearly ordered set $(I, <)$ there is a Morley sequence $(b_i : i \in I)$ in p. It is enough to obtain $(b_i : i \in I)$ as an A-indiscernible sequence with the same Ehrenfeucht–Mostowski set over A (the set of formulas over A realized by increasing finite tuples) as $(a_i : i < \omega)$.*

PROPOSITION 5.13. *If $a = (a_i : i < \omega)$ is a Morley sequence over A and $B \supseteq A$, there is a Morley sequence $b = (b_i : i < \omega)$ over B such that $a \equiv_A b$.*

PROOF. Let α be the length of each a_i, let $\kappa = |B| + |T| + |\alpha|$ and $\lambda = \beth_{(2^\kappa)^+}$. Extend a to an A-indiscernible sequence $(a_i : i < \lambda)$. It is also a Morley sequence over A. Construct inductively a sequence $(a'_i : i < \lambda)$ such that for all $i < \lambda$, $a_{<i} \equiv_A a'_{<i}$ and $a'_i \mathop{\smile\!\!\!\!\!\mid}_A Ba'_{<i}$. To obtain a'_i we choose some $f \in \mathrm{Aut}(\mathfrak{C}/A)$ such that $f(a_{<i}) = a'_{<i}$. Since $p(x) = \mathrm{tp}(a_i/Aa_{<i})$ does not fork over A, its conjugate $p^f(x) \in S(Aa'_{<i})$ does not fork over A and hence it has an extension $q(x) \in S(Ba'_{<i})$ which does not fork over A. We take as a'_i a realization of q. Then $a_{<i}a_i \equiv_A a'_{<i}a'_i$ and $a'_i \mathop{\smile\!\!\!\!\!\mid}_A Ba'_{<i}$. By Proposition 1.6 there is a B-indiscernible sequence $b = (b_i : i < \omega)$ such that for each $n < \omega$ there are $i_0 < \cdots < i_n < \lambda$ such that $b_0, \ldots, b_n \equiv_B a'_{i_0}, \ldots, a'_{i_n}$. Then

$$b_0, \ldots, b_n \equiv_A a'_{i_0}, \ldots, a'_{i_n} \equiv_A a_{i_0}, \ldots, a_{i_n} \equiv_A a_0, \ldots, a_n$$

and therefore $a \equiv_A b$. Since $a'_{i_n} \mathop{\smile\!\!\!\!\!\mid}_A Ba'_{<i_n}$, we see that $b_n \mathop{\smile\!\!\!\!\!\mid}_A Bb_{<n}$ and thus $b_n \mathop{\smile\!\!\!\!\!\mid}_B b_{<n}$. This shows that b is a Morley sequence over B. ⊣

LEMMA 5.14. *Let $(a_i : i \in I)$ be A-independent. If J, K are subsets of I such that $J < K$ (that is, $j < k$ for any $j \in J$, $k \in K$), then $\mathrm{tp}((a_i : i \in K)/A(a_i : i \in J))$ does not divide over A.*

PROOF. It can be assumed that K is finite. An induction on $|K|$ using Lemma 4.8 easily gives the result. ⊣

PROPOSITION 5.15. *Let T be simple, and let $\pi(x, y)$ be a set of formulas over A. Then $\pi(x, a)$ divides over A if and only if for every Morley sequence $(a_i : i < \omega)$ over A in $\mathrm{tp}(a/A)$, $\bigcup_{i<\omega} \pi(x, a_i)$ is inconsistent.*

PROOF. Without loss of generality $A = \emptyset$ and $\pi(x, y) = \{\varphi(x, y)\}$. Assume that $\varphi(x, a)$ divides over $\emptyset$ but for some infinite Morley sequence the inconsistency fails. Let $(I, <)$ be a linearly ordered set isomorphic to the reverse order of the cardinal $|T|^+$. By compactness there is a Morley sequence $a_I = (a_i : i \in I)$ in $\mathrm{tp}(a)$ such that $\{\varphi(x, a_i) : i \in I\}$ is consistent. Let c realize this type. By simplicity there is $J \subseteq I$ of cardinality at most $|T|$ such that $\mathrm{tp}(c/a_I)$ does not fork over $a_J = (a_i : i \in J)$. By choice of the order of I we can find $i \in I$ such that $i < J$. By Lemma 5.14 $\mathrm{tp}(a_J/a_i)$ does not divide over $\emptyset$. Since $\varphi(x, a_i)$ divides over $\emptyset$, by Proposition 4.9 it divides over a_J. But $\mathrm{tp}(c/a_I)$ contains $\varphi(x, a_i)$ and hence it divides (and forks) over a_J, a contradiction. ⊣

REMARK 5.16. *The previous result can be easily generalized to Morley sequences* $(a_i : i \in I)$ *for any infinite linear ordering* $(I, <)$.

PROOF. By compactness. ⊣

PROPOSITION 5.17. *Let* T *be simple. A partial type* $\pi(x)$ *divides over* A *if and only if it forks over* A.

PROOF. We may assume $\pi(x)$ is a formula $\varphi(x, a)$. Assume $\varphi(x, a)$ does not divide over A but it implies a disjunction $\varphi_1(x, a_1) \vee \cdots \vee \varphi_n(x, a_n)$ where every $\varphi_i(x, a_i)$ divides over A. Let $(a^j a_1^j \dots a_n^j : j < \omega)$ be a Morley sequence in $\mathrm{tp}(aa_1 \dots a_n/A)$. Then $(a^j : j < \omega)$ is an A-indiscernible sequence of realizations of $\mathrm{tp}(a/A)$. By definition of dividing, there exists a realization c of $\{\varphi(x, a^j) : j < \omega\}$. For every $j < \omega$ there is some i such that c realizes some $\varphi_i(x, a_i^j)$. By the pigeonhole principle, there is some i such that for an infinite subset $J \subseteq \omega$, c realizes every $\varphi_i(x, a_i^j)$ with $j \in J$. By indiscernibility, $\{\varphi_i(x, a_i^j) : j < \omega\}$ is consistent and then by Proposition 5.15 $\varphi_i(x, a_i)$ does not divide over A since $(a_i^j : j < \omega)$ is a Morley sequence in $\mathrm{tp}(a_i/A)$. ⊣

PROPOSITION 5.18 (Symmetry). *In a simple theory independence is a symmetric relation, i.e,* $A \mathop{\smile\!\!\!\!\!\!|}_C B$ *implies* $B \mathop{\smile\!\!\!\!\!\!|}_C A$.

PROOF. By Proposition 5.17, it is enough to prove that if $\mathrm{tp}(a/Cb)$ does not fork over C, then $\mathrm{tp}(b/Ca)$ does not divide over C. We may assume that $\mathrm{tp}(a/C)$ is not algebraic. By Lemma 5.11 there is a Morley sequence $I = (a_i : i < \omega)$ in $\mathrm{tp}(a/C)$ which is Cb-indiscernible and starts with $a_0 = a$. Let $\varphi(x, y, z) \in L$ be a formula and $c \in C$ such that $\models \varphi(a, b, c)$. We will show that $\varphi(a, y, c)$ does not divide over C. By indiscernibility of I over Cb we know that $\models \varphi(a_i, b, c)$ for all $i < \omega$. Hence $\{\varphi(a_i, y, c) : i < \omega\}$ is consistent. Since $(a_i c : i < \omega)$ is a Morley sequence in $\mathrm{tp}(ac/C)$, by Proposition 5.15 we conclude that $\varphi(a, y, c)$ does not divide over C. ⊣

PROPOSITION 5.19 (Transitivity). *In a simple theory independence is a transitive relation, i.e, whenever* $B \subseteq C \subseteq D$, $A \mathop{\smile\!\!\!\!\!\!|}_B C$ *and* $A \mathop{\smile\!\!\!\!\!\!|}_C D$, *then* $A \mathop{\smile\!\!\!\!\!\!|}_B D$.

PROOF. It is a direct consequence of propositions 5.18, 4.8, and 5.17. ⊣

PROPOSITION 5.20. *If* T *is simple, the independence relation has the following additional properties*:

1. *Pairs Lemma*: $ab \mathop{\smile\!\!\!\!\!\!|}_A B$ *if and only if* $a \mathop{\smile\!\!\!\!\!\!|}_A B$ *and* $b \mathop{\smile\!\!\!\!\!\!|}_{Aa} B$.
2. *Change of base*: *if* $ab \mathop{\smile\!\!\!\!\!\!|}_A B$, *then* $a \mathop{\smile\!\!\!\!\!\!|}_A b \Leftrightarrow a \mathop{\smile\!\!\!\!\!\!|}_{AB} b$.
3. $A \mathop{\smile\!\!\!\!\!\!|}_B \mathrm{acl}(B)$.
4. $A \mathop{\smile\!\!\!\!\!\!|}_B C \Leftrightarrow \mathrm{acl}(A) \mathop{\smile\!\!\!\!\!\!|}_B C \Leftrightarrow A \mathop{\smile\!\!\!\!\!\!|}_B \mathrm{acl}(C) \Leftrightarrow A \mathop{\smile\!\!\!\!\!\!|}_{\mathrm{acl}(B)} C$.

PROOF. It is an easy exercise. ⊣

COROLLARY 5.21. *Let* T *be simple. If* I *is a linearly ordered set and* $(a_i : i \in I)$ *is an* A-*independent sequence, then* $a_i \mathop{\smile\!\!\!\!\!\!|}_A \{a_j : j \neq i\}$ *for all* $i \in I$.

PROOF. By induction on n it is easy to show that $a_{i_{n+1}} \mathop{\smile\!\!\!\!\!\!\;|}_A a_{i_1}, \dots, a_{i_n}$ for all different $i_1, \dots, i_{n+1} \in I$. For the induction one uses symmetry and Lemma 4.8. ⊣

PROPOSITION 5.22. *Let T be simple, $p(x) \in S(A)$ and let $\pi(x)$ be a partial type over B. Then $p(x) \cup \pi(x)$ does not fork over A if and only if* $\mathrm{D}(p, \Delta, k) = \mathrm{D}(p \cup \pi, \Delta, k)$ *for all finite Δ, k.*

PROOF. We may assume $A \subseteq B$ and $\Delta = \{\varphi\}$. Direction from right to left follows from Lemma 4.14. Now assume $p \cup \pi$ is a nonforking extension of p and choose $q(x) \in S(B)$ a type which does not fork over A and extends $p \cup \pi$. We will check that $\mathrm{D}(q, \varphi, k) \geq \mathrm{D}(p, \varphi, k)$ for all φ, k. From this it will follow that $\mathrm{D}(p, \varphi, k) \leq \mathrm{D}(p \cup \pi, \varphi, k)$ for all φ, k. We freely use transitivity and symmetry of independence and also the fact that dividing and forking coincide. Let $n = \mathrm{D}(p, \varphi, k)$. There is a sequence $b_{<n} = (b_i : i < n)$ such that $p(x) \cup \{\varphi(x, b_i) : i < n\}$ is consistent and $\varphi(x, b_i)$ k-divides over $Ab_{<i}$ for all $i < n$. Let $a \models p(x) \cup \{\varphi(x, b_i) : i < n\}$, let $c \models q$ and let B' be such that $cB \equiv_A aB'$ and $B' \mathop{\smile\!\!\!\!\!\!\;|}_{Aa} b_{<n}$. Then, since $B \mathop{\smile\!\!\!\!\!\!\;|}_A c$, it follows that $B' \mathop{\smile\!\!\!\!\!\!\;|}_A a$ and by transitivity $B' \mathop{\smile\!\!\!\!\!\!\;|}_A b_{<n}$. By Proposition 4.9 $\varphi(x, b_i)$ k-divides over $B'b_{<i}$ for all $i < n$. For $q' = \mathrm{tp}(a/B')$ we then have $\mathrm{D}(q', \varphi, k) \geq n$. Since q is a conjugate of q', also $\mathrm{D}(q, \varphi, k) \geq n$. ⊣

COROLLARY 5.23. *Let T be simple. Assume $p(x) \in S(A)$ and let $\pi(x, y)$ be a set of formulas over A. Then*

$$\{a : p(x) \cup \pi(x, a) \textit{ does not fork over } A\}$$

is type-definable over A.

PROOF. For any $\varphi = \varphi(x, y) \in L$ and $k < \omega$, let $n_{\varphi,k} = \mathrm{D}(p(x), \varphi, k)$. By Proposition 5.22 we know that $p(x) \cup \pi(x, a)$ does not fork over A if and only if for all φ, k, $\mathrm{D}(p(x) \cup \pi(x, a), \varphi, k) \geq n_{\varphi,k}$, which (by Remark 3.3) can be expressed by a partial type over A. ⊣

COROLLARY 5.24. *Let T be simple and fix $p(x) \in S(A)$.*

1. *For any $n < \omega$*

   $$\{(a_1, \dots, a_n) : (a_1, \dots, a_n) \textit{ is } A\textit{-independent and } a_i \models p \textit{ for } i = 1, \dots, n\}$$

   is type-definable over A.
2. *For any linearly ordered set I,*

   $$\{(a_i : i \in I) : (a_i : i \in I) \textit{ is a Morley sequence in } p\}$$

   is type-definable over A.

PROOF. 2 follows from 1 and from the fact that indiscernibility over A is type-definable over A. As for 1, it can be proved as in Corollary 5.23, but now using Remark 3.14 instead of Remark 3.3: it is enough to express that each a_i realizes p and $\mathrm{D}(\mathrm{tp}(a_i/Aa_1, \dots, a_{i-1}), \varphi, k) \geq n_{\varphi,k}$ for all $\varphi = \varphi(x, y) \in L$, for all $k < \omega$, where $n_{\varphi,k} = \mathrm{D}(p(x), \varphi, k)$. ⊣

Chapter 6

THE LOCAL RANK $\mathrm{CB}_\Delta(\pi)$

DEFINITION 6.1. Let $\pi(x)$ be a set of formulas over the set A and let

$$\Delta = \{\varphi_1(x, y_1), \dots, \varphi_n(x, y_n)\}$$

be a finite set of formulas $\varphi_i(x, y_i) \in L$. Let $m < \omega$ be the length of the tuple of variables x. Since the restriction map $S_m(\mathfrak{C}) \to S_\Delta(\mathfrak{C})$ is closed and the class

$$X_{\pi,\Delta} = \{\mathfrak{p} \in S_\Delta(\mathfrak{C}) : \mathfrak{p}(x) \cup \pi(x) \text{ is consistent}\}$$

is the image of the closed class $\{\mathfrak{p} \in S_m(\mathfrak{C}) : \pi(x) \subseteq \mathfrak{p}\}$, $X_{\pi,\Delta}$ is closed in $S_\Delta(\mathfrak{C})$. We define the Δ-rank $\mathrm{CB}_\Delta(\pi)$ as the Cantor–Bendixson rank of $X_{\pi,\Delta}$ in $S_\Delta(\mathfrak{C})$ and the Δ-multiplicity $\mathrm{Mlt}_\Delta(\pi)$ as its Cantor–Bendixson degree.

LEMMA 6.2. *If $\pi_1(x) \vdash \pi_2(x)$, then $\mathrm{CB}_\Delta(\pi_1) \leq \mathrm{CB}_\Delta(\pi_2)$ and in case $\mathrm{CB}_\Delta(\pi_1) = \mathrm{CB}_\Delta(\pi_2)$, then $\mathrm{Mlt}_\Delta(\pi_1) \leq \mathrm{Mlt}_\Delta(\pi_2)$.*

PROOF. Clear, because if $X_{\pi_i,\Delta} = \{\mathfrak{p} \in S_\Delta(\mathfrak{C}) : \mathfrak{p} \text{ is consistent with } \pi_i\}$, then $X_{\pi_1,\Delta} \subseteq X_{\pi_2,\Delta}$. ⊣

REMARK 6.3. *For each $\Delta = \{\varphi_1(x, y_1), \dots, \varphi_n(x, y_n)\}$ where $\varphi_i(x, y_i) \in L$ for every i, there is a formula $\psi_\Delta(x; z) \in L$ such that for each partial type $\pi(x)$, $\mathrm{CB}_\Delta(\pi) = \mathrm{CB}_{\psi_\Delta}(\pi)$ and $\mathrm{Mlt}_\Delta(\pi) = \mathrm{Mlt}_{\psi_\Delta}(\pi)$.*

PROOF. By Lemma 3.4. ⊣

PROPOSITION 6.4. *Let $\psi(x)$ be a boolean combination of Δ-formulas.*

1. *$\mathrm{CB}_\Delta(\psi) \geq 0$ if and only if ψ is consistent.*
2. *$\mathrm{CB}_\Delta(\psi) \geq \alpha + 1$ if and only if there is a sequence $(\psi_i(x) : i < \omega)$ of pairwise contradictory boolean combinations $\psi_i(x)$ of Δ-formulas such that $\mathrm{CB}_\Delta(\psi(x) \wedge \psi_i(x)) \geq \alpha$ for all $i < \omega$.*
3. *$\mathrm{CB}_\Delta(\psi) \geq \alpha$ for limit α if and only if $\mathrm{CB}_\Delta(\psi) \geq \beta$ for all $\beta < \alpha$.*
4. *If $\mathrm{CB}_\Delta(\psi) = \alpha < \infty$, then $\mathrm{Mlt}_\Delta(\psi)$ is the largest $n < \omega$ for which there is a sequence $(\psi_i(x) : i < n)$ of pairwise contradictory boolean combinations $\psi_i(x)$ of Δ-formulas such that $\mathrm{CB}_\Delta(\psi(x) \wedge \psi_i(x)) \geq \alpha$ for all $i < n$.*

The formulas ψ_i in 2 and 4 can be chosen as explicitly contradictory conjunctions of Δ-formulas. Moreover in 2 we can fix some $\varphi \in \Delta$ such that each ψ_i is a conjunction of φ-formulas.

PROOF. Let $X_{\Delta,\psi}$ be the clopen subset of $S_\Delta(\mathfrak{C})$ consisting of all types $\mathfrak{p} \in S_\Delta(\mathfrak{C})$ such that $\mathfrak{p} \vdash \psi$. $\mathrm{CB}_\Delta(\psi)$ is the maximal Cantor–Bendixson rank (in $S_\Delta(\mathfrak{C})$) of the points in $X_{\Delta,\psi}$. Points 1 and 3 are clear. The proof of 4 is similar to the proof of 2, so we restrict ourselves to 2. Assume first there is a sequence $(\psi_i(x) : i < \omega)$ of pairwise contradictory boolean combinations ψ_i of Δ-formulas such that $\mathrm{CB}_\Delta(\psi \wedge \psi_i) \geq \alpha$ for each $i < \omega$. For each i choose some $\mathfrak{p}_i \in S_\Delta(\mathfrak{C})$ of Cantor–Bendixson rank $\geq \alpha$ such that $\mathfrak{p}_i \vdash \psi \wedge \psi_i$. Since the ψ_i are pairwise contradictory, all the $\mathfrak{p}_i$ are different. Since $X_{\Delta,\psi}$ contains infinitely many points of rank $\geq \alpha$, it contains some point of rank $\geq \alpha + 1$. Hence $\mathrm{CB}_\Delta(\psi) \geq \alpha + 1$.

For the other direction, assume $\mathrm{CB}_\Delta(\psi) \geq \alpha + 1$. Then $X_{\Delta,\psi}$ is an open set containing a point of rank $\geq \alpha + 1$. Thus the set Y_0 of points of $X_{\Delta,\psi}$ of rank $\geq \alpha$ is infinite. Clearly, for some Δ-formula θ there are points in Y_0 containing θ and points in Y_0 containing $\neg\theta$ and one of them, say the second one, is infinite. Let $\theta_0 = \theta$ and let Y_1 be the infinite subset of Y_0 consisting of all points containing $\neg\theta_0$. Now assume that Y_i, ψ_i are defined for all $i \leq n$, assume that the Y_i build a strictly descending chain of infinite sets, and assume that Y_{i+1} is the subset of Y_i consisting of all its points containing the Δ-formula $\neg\theta_i$. Again, there is some Δ-formula θ_{n+1} such that some points of Y_{n+1} contain θ_{n+1} and infinitely many points of Y_{n+1} contain $\neg\theta_{n+1}$. For some infinite subset $I \subseteq \omega$ there is a $\varphi \in \Delta$ such that for each $i \in I$, θ_i is a φ-formula. Without loss of generality, $I = \omega$. We then put $\psi_n = \theta_n \wedge \bigwedge_{i<n} \neg\theta_i$. ⊣

PROPOSITION 6.5. *Fix Δ and $\pi(x)$.*

1. *There is a boolean combination ψ of Δ-formulas such that $\pi(x) \vdash \psi(x)$,* $\mathrm{CB}_\Delta(\pi) = \mathrm{CB}_\Delta(\psi)$, *and* $\mathrm{Mlt}_\Delta(\pi) = \mathrm{Mlt}_\Delta(\psi)$.
2. *If $\pi(x)$ is over A, it can be extended to a complete type $p(x) \in S(A)$ such that* $\mathrm{CB}_\Delta(\pi) = \mathrm{CB}_\Delta(p)$.

PROOF. 1. Let $X = \{\mathfrak{p} \in S_\Delta(\mathfrak{C}) : \pi(x) \cup \mathfrak{p}(x) \text{ is consistent}\}$ be the closed set in $S_\Delta(\mathfrak{C})$ whose Cantor–Bendixson rank is $\mathrm{CB}_\Delta(\pi)$. For general topological reasons, there is a clopen set $U \supseteq X$ of the same Cantor–Bendixson rank and degree. The required formula $\psi(x)$ is the boolean combination of Δ-formulas such that $U = \{\mathfrak{p} \in S_\Delta(\mathfrak{C}) : \mathfrak{p} \vdash \psi\}$.

2. Take $\mathfrak{p}(x) \in S_\Delta(\mathfrak{C})$ consistent with $\pi(x)$ and of Cantor–Bendixson rank $\mathrm{CB}_\Delta(\pi)$ and take any extension $p(x) \in S(A)$ consistent with $\mathfrak{p}(x)$. Clearly $\mathrm{CB}_\Delta(p)$ is still the rank of $\mathfrak{p}$. ⊣

PROPOSITION 6.6. *The following are equivalent:*

1. *Every $\varphi(x, y) \in \Delta$ is stable.*
2. $\mathrm{CB}_\Delta(x = x) < \omega$.
3. $\mathrm{CB}_\Delta(x = x) < \infty$.

PROOF. Clearly 2 implies 3. Stability of every $\varphi \subset \Delta$ means that for each infinite set A, $|S_\Delta(A)| \leq |A|$. It is therefore equivalent to the stability of the formula ψ_Δ given by Lemma 3.4. Hence we may assume that $\Delta = \{\varphi\}$. By Proposition 2.6, the stability of φ is equivalent to the inconsistency of the set of formulas $\Gamma_\varphi(\omega)$, where for each ordinal α,

$$\Gamma_\varphi(\alpha) = \{\varphi(x_f, y_{f\restriction y})^{f(i)} : f \in 2^\alpha,\, i < \alpha\}$$

and where $\varphi^0 = \varphi$ and $\varphi^1 = \neg\varphi$.

$1 \Rightarrow 2$. Assume $\mathrm{CB}_\varphi(x = x) \geq \omega$. If $\psi(x)$ is a boolean combination of φ-formulas and $\mathrm{CB}_\varphi(\psi) \geq n + 1$ then for some a, $\mathrm{CB}_\varphi(\psi(x) \wedge \varphi(x, a)) \geq n$ and $\mathrm{CB}_\varphi(\psi \wedge \neg\varphi(x, a)) \geq n$. Since $\mathrm{CB}_\varphi(x = x) \geq \omega$ this can be used to construct a binary tree of parameters $(a_s : s \in 2^{<n})$ such that for each $s \in 2^n$ the branch $\{\varphi(x, a_{s\restriction i})^{s(i)} : i < n\}$ is consistent. This implies that $\Gamma_\varphi(n)$ is consistent. By compactness $\Gamma_\varphi(\omega)$ is consistent and hence φ is unstable.

$3 \Rightarrow 1$. Assume φ is unstable but $\mathrm{CB}_\varphi(x = x) < \infty$. Hence $\Gamma_\varphi(\omega)$ is consistent and we may find parameters $(a_s : s \in 2^{<\omega})$ such that for each $f \in 2^\omega$ the branch $\{\varphi(x, a_{f\restriction i})^{f(i)} : i < \omega\}$ is consistent. For any $s \in 2^n$, $n < \omega$, let

$$\psi_s(x) = \bigwedge_{i<n} \varphi(x, a_{s\restriction i})^{s(i)}$$

and choose s for which ψ_s has minimal CB_φ-rank and least multiplicity Mlt_φ among the formulas with the same rank. But $\psi_s(x)$ is equivalent to $(\psi_{s^\frown 0}(x) \vee \psi_{s^\frown 1}(x))$ and the formulas $\psi_{s^\frown 0}$, $\psi_{s^\frown 1}$ are incompatible. So one of them has smaller CB_φ-rank or they have the same rank and one has smaller φ-multiplicity, a contradiction. ⊣

REMARK 6.7. *Let $\varphi = \varphi(x, y) \in L$ be stable, let $\pi(x)$ be a partial type over A, and let $\mathfrak{p} \in S_\varphi(\mathfrak{C})$ be consistent with $\pi(x)$ and of Cantor–Bendixson rank $\mathrm{CB}_\varphi(\pi)$. Then $\mathfrak{p}$ is definable over* $\mathrm{acl}^{\mathrm{eq}}(A)$. *If* $\mathrm{Mlt}_\varphi(\pi) = 1$ *it is also A-definable.*

PROOF. By stability of φ, $\mathfrak{p}$ is definable (see Remark 2.7). All the A-conjugates of $\mathfrak{p}$ have Cantor–Bendixson rank $\mathrm{CB}_\varphi(\pi)$ and are consistent with $\pi(x)$; their number is bounded by $\mathrm{Mlt}_\varphi(\pi) < \omega$. Since $\mathfrak{p}$ has finitely many A-conjugates, by Proposition 1.11 $\mathfrak{p}$ is $\mathrm{acl}^{\mathrm{eq}}(A)$-definable. If $\mathrm{Mlt}_\varphi(\pi) = 1$, $\mathfrak{p}$ is A-invariant and therefore A-definable. ⊣

LEMMA 6.8. *Let $\varphi(x, y) \in L$ be stable. Let $\mathfrak{p}(x) \in S_\varphi(\mathfrak{C})$ and $\mathfrak{q}(y) \in S_{\varphi^{-1}}(\mathfrak{C})$ and let $d_\mathfrak{p} x\varphi(x, y)$ and $d_\mathfrak{q} y\varphi(x, y)$ be corresponding definitions of $\mathfrak{p}$ and $\mathfrak{q}$ which are boolean combinations of φ^{-1}-formulas and of φ-formulas respectively. Then $\mathfrak{q} \vdash d_\mathfrak{p} x\varphi(x, y)$ if and only if $\mathfrak{p} \vdash d_\mathfrak{q} y\varphi(x, y)$.*

PROOF. Let A be a set containing all the parameters of the formulas $d_\mathfrak{p} x\varphi(x, y)$ and $d_\mathfrak{q} y\varphi(x, y)$ defining respectively $\mathfrak{p}$ and $\mathfrak{q}$. Let $(a_n : n \in \omega)$ and $(b_n : n \in \omega)$ be sequences such that $a_n \models \mathfrak{p} \restriction A\{b_i : i < n\}$ and $b_n \models \mathfrak{q} \restriction A\{a_i : i \leq n\}$. If $\mathfrak{q} \vdash d_\mathfrak{p} x\varphi(x, y)$ and $\mathfrak{p} \nvdash d_\mathfrak{q} y\varphi(x, y)$, we would have $\models \varphi(a_m, b_n)$ if and only if $m > n$, and therefore $\varphi(x, y)$ would have the order property. ⊣

DEFINITION 6.9. A *generalized φ-formula over A* is a formula over A which is equivalent to a boolean combination of φ-formulas, possibly with parameters not in A. A *generalized φ-type over A* is a consistent set of generalized φ-formulas over A. A generalized φ-type $p(x)$ over A is *complete* if for every generalized φ-formula $\psi(x)$ over A, $\psi(x) \in p$ or $\neg\psi(x) \in p$. The set of all complete generalized φ-types over A is $S^*_\varphi(A)$. The generalized φ-type of a over A is $\mathrm{tp}^*_\varphi(a/A)$, the complete generalized φ-type which consists of all generalized φ-formulas over A realized by a.

As in the cases of $S_\varphi(A)$ and $S(A)$, we can treat $S^*_\varphi(A)$ as a boolean topological space with basis of clopen sets given by $[\psi(x)] = \{p(x) \in S^*_\varphi(A) : \psi \in p\}$ for every generalized φ-formula $\psi(x)$ over A. The restriction mappings $S(A) \to S^*_\varphi(A)$ and $S^*_\varphi(A) \to S_\varphi(A)$ are surjective and continuous.

REMARK 6.10. *For any model M, every generalized φ-formula over M is equivalent to a boolean combination of φ-formulas over M. Hence, every complete φ-type over M can be uniquely extended to a complete generalized φ-type over M and the restriction map $S^*_\varphi(M) \to S_\varphi(M)$ is an homeomorphism.*

PROOF. Assume $\psi(x) \in L(M)$ is equivalent to a boolean combination $\theta(x, c_1, \ldots, c_n)$ of φ-formulas and $\theta(x, x_1, \ldots, x_n) \in L$. Then

$$M \models \exists x_1 \ldots x_n \forall x(\psi(x) \leftrightarrow \theta(x, x_1, \ldots, x_n))$$

and therefore we can find $c'_1, \ldots, c'_n \in M$ such that $\psi(x) \equiv \theta(x, c'_1, \ldots, c'_n)$. ⊣

REMARK 6.11. *If $\varphi(x, y)$ is stable and $\mathfrak{p} \in S_\varphi(\mathfrak{C})$ is definable over A, then it is definable by a generalized φ^{-1}-formula over A.*

PROOF. By Remark 2.13. ⊣

LEMMA 6.12. *Assume $p(x) \in S^*_\varphi(A)$ and $q(x) \in S^*_\varphi(\mathrm{acl}(A))$ is an extension of p. Then $q(x)$ is consistent with any extension $p'(x) \in S(A)$ of p.*

PROOF. Suppose $p'(x) \in S(A)$ extends $p(x)$ and is inconsistent with $q(x)$. Then for some $\psi(x) \in q(x)$, $p'(x) \vdash \neg\psi(x)$. If $\psi'(x)$ is an A-conjugate of $\psi(x)$, then also $p'(x) \vdash \neg\psi'(x)$. Since $\psi(x) \in L(\mathrm{acl}(A))$, it has only finitely many A-conjugates and we can form their disjunction $\theta(x)$. Since $\theta(x)$ is (equivalent to) a generalized φ-formula over A and $p'(x) \vdash \neg\theta(x)$ we conclude $\neg\psi(x) \in q$, a contradiction. ⊣

PROPOSITION 6.13. *Let φ be stable.*

1. *If $p(x) \in S_\varphi(M)$, then there is a unique $\mathfrak{p}(x) \in S_\varphi(\mathfrak{C})$ extending p which is definable over M and hence* $\mathrm{Mlt}_\varphi(p) = 1$.
2. *If $A = \mathrm{acl}^{\mathrm{eq}}(A)$ and $p(x) \in S^*_\varphi(A)$, there is a unique $\mathfrak{p}(x) \in S_\varphi(\mathfrak{C})$ consistent with p and definable over A and hence* $\mathrm{Mlt}_\varphi(p) = 1$.

PROOF. 1 follows from Lemma 2.4, but also can be considered a particular case of 2.

2. Existence follows from Remark 6.7. For uniqueness, let $\mathfrak{p}_1, \mathfrak{p}_2 \in S_\varphi(\mathfrak{C})$ be global φ-types consistent with p and A-definable. By Remark 6.11 there are corresponding definitions $\psi_i(y) \in L(A)$ $(i = 1, 2)$ which are generalized φ^{-1}-formulas over A. Recall that φ^{-1} is also stable. Let b be a tuple of the same length as y and let us choose (by Remark 6.7) a global type $\mathfrak{q}(y) \in S_{\varphi^{-1}}(\mathfrak{C})$ consistent with $\mathrm{tp}(b/A)$ and definable over A by a formula $\theta(x) \in L(A)$. By Remark 6.11 again, we may assume $\theta(x)$ is a generalized φ-formula over A. We apply now Lemma 6.8 with $\psi_i'(y) = d_{\mathfrak{p}_i}x\varphi(x, y)$ and $\theta(x) = d_{\mathfrak{q}}y\varphi(x, y)$ obtaining:

$\varphi(x, b) \in \mathfrak{p}_i \Leftrightarrow \models \psi_i(b)$	because ψ_i defines $\mathfrak{p}_i$
$\Leftrightarrow \psi_i(y) \in \mathrm{tp}(b/A)$	because $\psi_i(y) \in L(A)$
$\Leftrightarrow \mathfrak{q} \vdash \psi_i(y)$	because $\mathfrak{q}(y) \cup \mathrm{tp}(b/A)$ is consistent and ψ_i is equivalent to a boolean combination of φ^{-1}-formulas
$\Leftrightarrow \mathfrak{p}_i \vdash \theta(x)$	by Lemma 6.8
$\Leftrightarrow \theta \in p$	since $p(x) \cup \mathfrak{p}_i$ is consistent and θ is a generalized φ-formula over A.

This shows that $\mathfrak{p}_1 = \mathfrak{p}_2$. ⊣

COROLLARY 6.14. *Let $\varphi = \varphi(x, y) \in L$ be stable and let $p(x) \in S^*_\varphi(A)$. Every two types $\mathfrak{p}(x), \mathfrak{q}(x) \in S_\varphi(\mathfrak{C})$ consistent with $p(x)$ and definable over* $\mathrm{acl}^{\mathrm{eq}}(A)$ *are A-conjugate.*

PROOF. Let $\mathfrak{p}, \mathfrak{q}$ be two such types. They can be considered complete generalized φ-types over $\mathfrak{C}$. Let $p_1, q_1 \in S^*_\varphi(\mathrm{acl}^{\mathrm{eq}}(A))$ be their corresponding restrictions. By Lemma 6.12 (applied in T^{eq}) there is some $p'(x) \in S(A)$ extending p which is consistent with $p_1(x)$ and is consistent with $q_1(x)$. Now let $p_1', q_1' \in S(\mathrm{acl}^{\mathrm{eq}}(A))$ be extensions of $p' \cup p_1$ and of $p' \cup p_2$ respectively. Clearly there is some $f \in \mathrm{Aut}(\mathfrak{C}/A)$ sending p_1' to q_1'. Then $\mathfrak{q}$ and $\mathfrak{p}^f$ are $\mathrm{acl}^{\mathrm{eq}}(A)$-definable and consistent with q_1. By Proposition 6.13 $\mathfrak{p}^f = \mathfrak{q}$. ⊣

COROLLARY 6.15. *Let $\varphi = \varphi(x, y) \in L$ be stable, let $p(x) \in S^*_\varphi(A)$. For every $\mathfrak{p}(x) \in S_\varphi(\mathfrak{C})$ consistent with p, the following are equivalent:*

1. *$\mathfrak{p}$ is definable over* $\mathrm{acl}^{\mathrm{eq}}(A)$.
2. *$\mathfrak{p}$ is a point of Cantor–Bendixson rank* $\mathrm{CB}_\varphi(p)$.

In case $\mathrm{Mlt}_\varphi(p) = 1$ *there is a unique such $\mathfrak{p} \in S_\varphi$ and it is in fact A-definable.*

PROOF. Let $X_{p,\varphi} \subseteq S_\varphi(\mathfrak{C})$ be the class of all global φ-types consistent with p. By Remark 6.7 we know that all types in $X_{p,\varphi}$ of rank $\mathrm{CB}_\varphi(p)$ are definable over $\mathrm{acl}^{\mathrm{eq}}(A)$. Now let $\mathfrak{p}, \mathfrak{q} \in X_{p,\varphi}$ be such that $\mathfrak{p}$ is $\mathrm{acl}^{\mathrm{eq}}(A)$-definable and $\mathfrak{q}$ has Cantor–Bendixson rank $\mathrm{CB}_\varphi(p)$. By Corollary 6.14 they are A-conjugate and therefore $\mathfrak{p}$ has also rank $\mathrm{CB}_\varphi(p)$ in $X_{p,\varphi}$. ⊣

Chapter 7

HEIRS AND COHEIRS

DEFINITION 7.1. Let $M \subseteq A$ and $p(x) \in S(A)$. We say that p is an *heir* of $p \restriction M$ or that p *inherits* from M if for every $\varphi(x, y) \in L(M)$ if $\varphi(x, a) \in p$ for some tuple $a \in A$, then $\varphi(x, m) \in p$ for some tuple $m \in M$. We say that p is a *coheir* of $p \restriction M$ or that p *coinherits* from M if p is finitely satisfiable in M. The same definitions apply to global types, i.e, to the case $A = \mathfrak{C}$. These definitions also make sense for types in infinitely many variables.

REMARK 7.2. *$\mathrm{tp}(a/Mb)$ inherits from M if and only if $\mathrm{tp}(b/Ma)$ coinherits from M.*

PROOF. It is just a matter of writing down the definitions. ⊣

LEMMA 7.3. 1. *If $p(x) \in S(M)$, then p inherits and coinherits from M.*
2. *If $M \subseteq A$ and $p(x) \in S(A)$ coinherits from M, then for every $B \supseteq A$ there is some $q(x) \in S(B)$ such that $p \subseteq q$ and q coinherits from M.*
3. *If $M \subseteq A$ and $p(x) \in S(A)$ inherits from M, then for every $B \supseteq A$ there is some $q(x) \in S(B)$ such that $p \subseteq q$ and q inherits from M.*

PROOF. 1 is clear. For 2 it is enough to check the consistency of the following set of formulas

$$p(x) \cup \{\neg\varphi(x) : \varphi(x) \in L(B) \text{ is not satisfiable in } M\}.$$

3. In this case it suffices to prove that the following set of formulas is consistent

$$p(x) \cup \{\neg\varphi(x, b) : \varphi(x, y) \in L(M),\ b \in B \text{ and there is no } m \in M \text{ such that } \varphi(x, m) \in p \restriction M\}.$$ ⊣

DEFINITION 7.4. Let $p(x) \in S(B)$ and let $A \subseteq B$. We say that p *splits* over A if for some $\varphi(x, y) \in L(A)$ there are $a, b \in B$ such that $a \equiv_A b$, $\varphi(x, a) \in p$ and $\neg\varphi(x, b) \in p$. This applies also to the case $B = \mathfrak{C}$. Note that the same notion is defined if one requires $\varphi(x, y) \in L$. If $\mathfrak{p} \in S(\mathfrak{C})$, then clearly $\mathfrak{p}$ does not split over A if and only if $\mathfrak{p}^f = \mathfrak{p}$ for each $f \in \mathrm{Aut}(\mathfrak{C}/A)$. If moreover $A = M$, a global nonsplitting extension is also sometimes called an *M-special* extension.

PROPOSITION 7.5. 1. *The number of global nonsplitting extensions of a finitary type* $p \in S(A)$ *is* $\leq 2^{2^{|A|+|T|}}$.

2. *Let* $\mathfrak{p}$ *be a global nonsplitting extension of* $p \in S(A)$. *If the sequence* $(a_i : i < \alpha)$ *is constructed in such a way that for all* $i < \alpha$,

$$a_i \models \mathfrak{p} \upharpoonright Aa_{<i},$$

then it is A-indiscernible.

PROOF. 1. For each $\varphi(x, y) \in L$, the number of restrictions $\mathfrak{p} \upharpoonright \varphi$ for types $\mathfrak{p}(x) \in S(\mathfrak{C})$ which do not split over A is bounded by the number of sets of types $\text{tp}(a/A)$ of tuples $a \in \mathfrak{C}$ of the length of y. The number of these types is $\leq 2^{|A|+|T|}$ and therefore the number of sets of types is $\leq 2^{2^{|A|+|T|}}$.

2. By induction on n we show that for all $i_0 < \cdots < i_n < \alpha$,

$$a_0, \ldots, a_n \equiv_A a_{i_0}, \ldots, a_{i_n}.$$

This is clear for $n = 0$ since $\text{tp}(a_{i_0}/A) = p = \text{tp}(a_0/A)$. Consider the case $n + 1$. Let $\varphi(x_0, \ldots, x_{n+1}) \in L(A)$, and let $i_0 < \cdots < i_{n+1} < \alpha$. By the induction hypothesis $a_0, \ldots, a_n \equiv_A a_{i_0}, \ldots, a_{i_n}$. Since $\mathfrak{p}$ does not split over A, $\varphi(a_0, \ldots, a_n, x) \in \mathfrak{p}$ if and only if $\varphi(a_{i_0}, \ldots, a_{i_n}, x) \in \mathfrak{p}$. Since $a_{n+1} \models \mathfrak{p} \upharpoonright Aa_0, \ldots, a_n$ and $a_{i_{n+1}} \models \mathfrak{p} \upharpoonright Aa_{i_0}, \ldots, a_{i_n}$, we conclude that $\models \varphi(a_0, \ldots, a_n, a_{n+1})$ if and only if $\models \varphi(a_{i_0}, \ldots, a_{i_n}, a_{i_{n+1}})$. ⊣

PROPOSITION 7.6. 1. *Coheirs are nonsplitting extensions.*

2. *If* $\mathfrak{p} \in S(\mathfrak{C})$ *does not split over* A, *then it does not fork over* A.
3. *Coheirs are nonforking extensions.*
4. *If* $p(x) \in S(M)$ *is definable, then its* M*-definable extension over* $A \supseteq M$ *is the only heir of* p *over* A.
5. *In a simple theory, heirs are nonforking extensions.*

PROOF. 1. Suppose $p(x) \in S(A)$ coinherits from $M \subseteq A$, $a, b \in A$, $a \equiv_M b$, $\varphi(x, y) \in L(M)$ and $\varphi(x, a) \in p$ while $\neg\varphi(x, b) \in p$. Then some $c \in M$ satisfies $\varphi(x, a) \wedge \neg\varphi(x, b)$, which is impossible if $a \equiv_M b$.

2. For a global type forking and dividing is the same. Let $\varphi(x, y) \in L$. If $\varphi(x, a) \in \mathfrak{p}$ and $a_i \equiv_A a$ for each $i < \omega$, then $\varphi(x, a_i) \in \mathfrak{p}$ for each $i < \omega$ and hence $\{\varphi(x, a_i) : i < \omega\}$ is consistent.

3. This can be proved using points 1 and 2 and the extension property of coheirs, but it is also an immediate consequence of the definition of forking as indicated in Remark 4.4.

4. Let $p \in S(M)$ be definable. By the uniqueness of the M-definable extension, we only need to show that heirs are M-definable. Let $q \in S(A)$ be an heir of p, let $\varphi(x, y) \in L$ and $d_p x\varphi(x, y) \in L(M)$ a definition of $p \upharpoonright \varphi$. We show that it is also a definition of $q \upharpoonright \varphi$. If it is not a definition, then for some $a \in A$, $\neg(d_p x\varphi(x, a) \leftrightarrow \varphi(x, a)) \in q$ and therefore for some $a' \in M$, $\neg(d_p x\varphi(x, a') \leftrightarrow \varphi(x, a')) \in p$ contradicting the fact that $d_p x\varphi(x, y)$ defines $p \upharpoonright \varphi$.

5. Let T be simple and assume $p(x) \in S(A)$ inherits from $M \subseteq A$. Let $a \in A$ be a tuple and let $b \models p$. We want to show that $b \perp_M a$. By Remark 7.2 $\text{tp}(a/Mb)$ coinherits from M and by point 3, $a \perp_M b$. The result then follows by symmetry of independence. ⊣

DEFINITION 7.7. Given $p(x) \in S(M)$, by (M, dp) we refer to the expansion of M to language $L \cup \{R_\varphi : \varphi \in L\}$ where for every $\varphi = \varphi(x, y) \in L$, if $y = y_1, \dots, y_n$ then R_φ is n-ary and it is interpreted as $\{a \in M : \varphi(x, a) \in p\}$. Let $M \preceq N$ and $p(x) \subseteq q(x) \in S(N)$. We say that q is a *strong heir* of p if $(M, dp) \preceq (N, dq)$. This also makes sense when $N = \mathfrak{C}$.

REMARK 7.8. 1. *Strong heirs are heirs.*

2. *If $(M, dp) \preceq N'$ and $N = N' \restriction L$, then for some $q(x) \in S(N)$, $p \subseteq q$ and $N' = (N, dq)$.*
3. *Any strong heir of a nondefinable type is again nondefinable.*

PROOF. 1 and 2 are easy. For 3, let $q(x) \in S(N)$ be a strong heir of p, and assume $\varphi(x, y) \in L$, $\psi(y, z) \in L$, $n \in N$ and $\psi(y, n)$ defines $q \restriction \varphi$. Then $(N, dq) \models \exists z \forall y(\psi(y, z) \leftrightarrow R_\varphi(y))$. Since $(M, dp) \preceq (N, dq)$, for some $m \in M$, $(M, dp) \models \forall y(\psi(y, m) \leftrightarrow R_\varphi(y))$. Then $\psi(y, m)$ defines $p \restriction \varphi$. ⊣

PROPOSITION 7.9. *If $p(x) \in S(M)$ is not definable, then $p(x)$ has unboundedly many (nondefinable) strong heirs over $\mathfrak{C}$.*

PROOF. We first show that $p(x)$ has two strong heirs over some $N \succeq M$. Since p is not definable, (M, dp) is not a definable expansion of M. By Svenonius's Theorem (see, for instance, Théorème 9.02 in [36]), there is some $N' \succeq (M, dp)$ having some $f \in \text{Aut}(N' \restriction L/M)$ such that $f \notin \text{Aut}(N')$. Let $N = N' \restriction L$. Then for some $q(x) \in S(N)$, $N' = (N, dq)$ and $q^f \neq q$. Clearly q and q^f are two strong heirs of p. Since a strong heir of a nondefinable type is again nondefinable, we can iterate this procedure (taking unions at limits) obtaining for each cardinal κ a family $(p_i(x) : i < \kappa)$ of strong heirs $p_i \in S(M_i)$ of p such that $p_i \cup p_j$ is inconsistent if $i \neq j$. Clearly, each p_i can be extended to a type $\mathfrak{p}_i$ over $\mathfrak{C}$ which is a strong heir of p_i and therefore also of p. ⊣

DEFINITION 7.10. A *coheir sequence* over A is a sequence $(a_i : i < \alpha)$ such that for some $M \subseteq A$, for all $i < j < \alpha$, $\text{tp}(a_i/Aa_{<i}) = \text{tp}(a_j/Aa_{<i})$ and $\text{tp}(a_j/Aa_{<j})$ coinherits from M.

REMARK 7.11. 1. *Any coheir sequence over A is a Morley sequence over A.*

2. *For any $p(x) \in S(A)$ which coinherits from $M \subseteq A$ there is a coheir sequence $(a_i : i < \alpha)$ over A.*

PROOF. 1. Let $p_i = \text{tp}(a_i/Aa_{<i})$ and $p_\alpha = \bigcup_{i<\alpha} p_i$. Clearly $a_i \models p_\alpha \restriction Aa_{<i}$ and p_α coinherits from M. By point 3 of Proposition 7.6 the sequence is A-independent. By point 1 of Proposition 7.6 and point 2 of Proposition 7.5, it is A-indiscernible.

2. Choose an extension $\mathfrak{p} \in S(\mathfrak{C})$ of p which coinherits from M and choose $a_i \models \mathfrak{p} \restriction Aa_{<i}$. ⊣

PROPOSITION 7.12. *Let $\mathfrak{p} \in S(\mathfrak{C})$ be definable. Then $\mathfrak{p}$ does not split over A if and only if it is A-definable.*

PROOF. Let $\mathfrak{p}$ be definable and assume it does not split over A. For each $\varphi(x, y) \in L$, $\{a : \varphi(x, a) \in \mathfrak{p}\}$ is definable and A-invariant, and therefore it is definable over A. The other direction is immediate. ⊣

COROLLARY 7.13. *If $\mathfrak{p} \in S(\mathfrak{C})$ is definable over A, then $\mathfrak{p}$ does not fork over A.*

PROOF. It is a consequence of Proposition 7.12 and point 2 of Proposition 7.6. ⊣

COROLLARY 7.14. *Let $p(x) \in S(M)$ and assume every complete extension of p is definable. Then an extension $q(x)$ of p is an heir of p if and only if it is a coheir of p if and only if it is M-definable.*

PROOF. The equivalence of M-definability and heir is given by point 4 in Proposition 7.6. For the rest, by points 2 and 3 of Lemma 7.3 it is enough to check the result in the case of a global extension $\mathfrak{p} \in S(\mathfrak{C})$ of p. Then we can apply Proposition 7.12 and point 1 of Proposition 7.6 to prove that coheirs are heirs. The uniqueness of heirs (point 4 in Proposition 7.6) shows then that heirs are coheirs. ⊣

COROLLARY 7.15. *T is stable if and only if in T all heirs are coheirs.*

PROOF. If T is stable, Corollary 7.14 establishes that heirs are coheirs. If T is not stable, there is some $p(x) \in S(M)$ not definable. By Proposition 7.9 p has unboundedly many heirs. Since coheirs do not split, by Proposition 7.5 the number of coheirs of p is bounded by $2^{2^{|M|+|T|}}$. Hence some heir is not a coheir. ⊣

COROLLARY 7.16. *The following are equivalent:*

1. *T is stable.*
2. *Each type $p(x) \in S(M)$ has a unique heir over any $A \supseteq M$.*
3. *Each type $p(x) \in S(M)$ has a bounded number of heirs over any $A \supseteq M$.*

PROOF. If T is stable, point 4 of Proposition 7.6 shows that p has a unique heir. If T is not stable, there is some nondefinable $p(x) \in S(M)$. By Proposition 7.9 p has unboundedly many strong heirs over $\mathfrak{C}$. Clearly, for each $A \supseteq M$ for any strong heir $\mathfrak{p} \in S(\mathfrak{C})$ of p, $\mathfrak{p} \restriction A$ is an heir of p. ⊣

Chapter 8

STABLE FORKING

PROPOSITION 8.1. *Let $\Delta = \{\varphi_i(x, y_i) : i < n\}$ be a set of stable formulas. A type $\mathfrak{p} \in S_\Delta(\mathfrak{C})$ is definable over a model M if and only if it is finitely satisfiable in M. In fact, if $\mathfrak{p}$ is M-definable and it is consistent with a partial type $\pi(x)$ over M, then $\pi(x) \cup \mathfrak{p}(x)$ is finitely satisfiable in M.*

PROOF. We may assume $\Delta = \{\varphi(x, y)\}$. Let $\mathfrak{p}$ be M-definable and let us choose by Remark 2.13 a definition $d_\mathfrak{p} x\varphi(x, y)$, which is a boolean combination of φ^{-1}-formulas over M. Let

$$\varphi(x, a_1), \ldots, \varphi(x, a_n), \neg\varphi(x, b_1), \ldots, \neg\varphi(x, b_m)$$

be formulas in $\mathfrak{p}$. For $1 \leq i \leq n$ and $1 \leq j \leq m$, let $q_i = \mathrm{tp}_{\varphi^{-1}}(a_i/M)$ and $r_j = \mathrm{tp}_{\varphi^{-1}}(b_j/M)$. Again by Remark 2.13 there are $\mathfrak{q}_i \in S_{\varphi^{-1}}(\mathfrak{C})$ and $\mathfrak{r}_j \in S_{\varphi^{-1}}(\mathfrak{C})$ extending q_i and r_j respectively and having definitions $d_{\mathfrak{q}_i} y\varphi(x, y)$ and $d_{\mathfrak{r}_j} y\varphi(x, y)$ which are boolean combinations of φ-formulas over M. Then $\models d_\mathfrak{p} x\varphi(x, a_i)$ and $\models \neg d_\mathfrak{p} x\varphi(x, b_j)$ and hence $\mathfrak{q}_i \vdash d_\mathfrak{p} x\varphi(x, y)$ and $\mathfrak{r}_j \vdash \neg d_\mathfrak{p} x\varphi(x, y)$. By Lemma 6.8, $\mathfrak{p} \vdash d_{\mathfrak{q}_i} y\varphi(x, y)$ and $\mathfrak{p} \vdash \neg d_{\mathfrak{r}_j} y\varphi(x, y)$. Since these are formulas over M, for some $c \in M$, $\models d_{\mathfrak{q}_i} y\varphi(c, y)$ and $\models \neg d_{\mathfrak{r}_j} y\varphi(c, y)$ for all i, j. Then $\models \varphi(c, a_i)$ and $\models \neg\varphi(c, b_j)$ for all i, j. Clearly this c can also be found realizing additionally a given finite subset of $\pi(x)$.

For the other direction, let us assume $dx\varphi(x, y)$ is a definition of $\mathfrak{p}$ which is not equivalent to a formula over M. Then we can find b, c such that $b \equiv_M c$ and $\models dx\varphi(x, b)$ but $\models \neg dx\varphi(x, c)$. In this case $\varphi(x, b) \in \mathfrak{p}$ and $\neg\varphi(x, c) \in \mathfrak{p}$ but there is no $a \in M$ such that $\models \varphi(a, b) \wedge \neg\varphi(a, c)$. Hence $\mathfrak{p}$ is not finitely satisfiable in M. ⊣

PROPOSITION 8.2. *Let $\varphi(x, y) \in L$ be stable, let $\mathfrak{p}(x) \in S_\varphi(\mathfrak{C})$ and assume $\mathfrak{p}$ is definable over M and it is consistent with $\pi(x)$, a partial type over M. For some $q(x) \in S(M)$ extending $\pi(x) \cup \mathfrak{p} \restriction M$ there is a Morley sequence $(c_i : i < \omega)$ in q such that $\mathfrak{p}$ is definable by a positive boolean combination of the formulas $\varphi(c_i, y)$.*

PROOF. By Proposition 8.1 $\pi(x) \cup \mathfrak{p}(x)$ is finitely satisfiable in M. It is easy to check the consistency of

$$\pi(x) \cup \mathfrak{p}(x) \cup \{\neg\psi(x) : \psi(x) \in L(\mathfrak{C}) \text{ is not satisfiable in } M\}.$$

Let $\mathfrak{q} \in S(\mathfrak{C})$ be an extension of this set of formulas. Clearly $\mathfrak{q}$ coinherits from M and $\mathfrak{q} \upharpoonright \varphi = \mathfrak{p}$. We claim that for some $n < \omega$ there is a sequence $(c_i : i < n)$ such that $c_i \models \mathfrak{q} \upharpoonright Mc_{<i}$ and $\mathfrak{p}$ is definable by a positive boolean combination of the formulas $\varphi(c_i, y)$. Note that if this is the case we can complete the sequence to $(c_i : i < \omega)$, a coheir sequence over M of realizations of $\mathfrak{p} \upharpoonright M$. By Remark 7.11 it is a Morley sequence over M (in $q = \mathfrak{q} \upharpoonright M$). Let us assume that there is no such sequence $(c_i : i < n)$. We proceed as in the proof of Lemma 2.10 obtaining a_i, b_i, c_i such that $\varphi(x, a_i) \in \mathfrak{p}$, $\neg\varphi(x, b_i) \in \mathfrak{p}$, $\models \varphi(c_j, a_i) \to \varphi(c_j, b_i)$ for all $j < i$ and

$$c_i \models \mathfrak{q} \upharpoonright Ma_{\leq i}b_{\leq i}c_{<i}.$$

As in the proof of Lemma 2.10, this implies that $\varphi(x, y)$ has the order property and is, therefore, unstable. ⊣

PROPOSITION 8.3. *Let $\varphi(x, y) \in L$ be stable. Given A and a, let us fix $\mathfrak{q}(y) \in S_{\varphi^{-1}}(\mathfrak{C})$, the unique φ^{-1}-type over $\mathfrak{C}$ consistent with* $\mathrm{tp}(a/\mathrm{acl}^{\mathrm{eq}}(A))$ *and definable over* $\mathrm{acl}^{\mathrm{eq}}(A)$. *Fix, a definition $d_{\mathfrak{q}} y\varphi(x, y)$ of $\mathfrak{q}$, as in Proposition* 8.2, *which is equivalent to a formula over* $\mathrm{acl}^{\mathrm{eq}}(A)$ *and it is a positive boolean combination of formulas $\varphi(x, c_i)$ where $(c_i : i < \omega)$ is an* $\mathrm{acl}^{\mathrm{eq}}(A)$*-indiscernible sequence of realizations c_i of* $\mathrm{tp}(a/\mathrm{acl}^{\mathrm{eq}}(A))$. *Let $\sigma(x)$ be the (finite) disjunction of all A-conjugates of $d_{\mathfrak{q}} y\varphi(x, y)$. For any partial type $\pi(x)$ over A, the following are equivalent.*

1. *$\varphi(x, a) \in \mathfrak{p}$ for some $\mathfrak{p} \in S_\varphi(\mathfrak{C})$ definable over* $\mathrm{acl}^{\mathrm{eq}}(A)$ *and consistent with $\pi(x)$.*
2. *$\pi(x) \cup \{\varphi(x, a)\}$ is finitely satisfiable in every model $M \supseteq A$.*
3. *$\pi(x) \cup \{\varphi(x, a)\}$ does not divide over A.*
4. *Every set of* $\mathrm{acl}^{\mathrm{eq}}(A)$*-conjugates of $\varphi(x, a)$ is consistent with $\pi(x)$.*
5. *$d_{\mathfrak{q}} y\varphi(x, y)$ is consistent with $\pi(x)$.*
6. *$\sigma(x)$ is consistent with $\pi(x)$.*
7. *Some positive boolean combination of A-conjugates of $\varphi(x, a)$ is equivalent to a formula over A consistent with $\pi(x)$.*

PROOF. $1 \Rightarrow 2$ follows directly from Proposition 8.1.

$2 \Rightarrow 3$. Let $\psi(x)$ be a finite conjunction of formulas in π and let $(a_i : i < \omega)$ be an A-indiscernible sequence starting with $a = a_0$. By Corollary 1.7 $(a_i : i < \omega)$ is indiscernible over some model $M \supseteq A$. There is some $c \in M$ such that $\models \varphi(c, a) \wedge \psi(c)$. By indiscernibility $\models \varphi(c, a_i)$ for every $i < \omega$. Therefore $\{\varphi(x, a_i) \wedge \psi(x) : i < \omega\}$ is consistent and $\varphi(x, a) \wedge \psi(x)$ does not divide over A.

$1 \Rightarrow 4$. Any $\mathrm{acl}^{\mathrm{eq}}(A)$-conjugate of $\varphi(x, a)$ is in $\mathfrak{p}$.

$3 \Rightarrow 5$. Since the sequence of parameters $(c_i : i < \omega)$ builds an indiscernible sequence over A and $a \equiv_A c_i$, $\pi(x) \cup \{\varphi(x, c_i) : i < \omega\}$ is consistent. Any positive boolean combination of the formulas $\varphi(x, c_i)$ is therefore consistent with π.

$4 \Rightarrow 5$. The same reason, since in fact every $\varphi(x, c_i)$ is an $\mathrm{acl}^{\mathrm{eq}}(A)$-conjugate of $\varphi(x, a)$.

$5 \Rightarrow 6$. Clear by construction of σ.

$6 \Rightarrow 7$. $\sigma(x)$ satisfies the requirements in 7.

$7 \Rightarrow 1$. Let $\sigma'(x)$ be a positive boolean combination of A-conjugates of $\varphi(x, a)$ which is equivalent to a formula over A and is consistent with π. By Remark 6.7 there is $\mathfrak{p} \in S_\varphi(\mathfrak{C})$ definable over $\mathrm{acl}^{\mathrm{eq}}(A)$ and consistent with $\pi(x) \cup \{\sigma'(x)\}$. Since $\sigma'(x)$ is a disjunction of conjunctions of A-conjugates of $\varphi(x, a)$, some A-conjugate of $\varphi(x, a)$ appears in $\mathfrak{p}$. By conjugation over A, there is also some $\mathfrak{p}' \in S_\varphi(\mathfrak{C})$ definable over $\mathrm{acl}^{\mathrm{eq}}(A)$ and consistent with $\pi(x)$ such that $\varphi(x, a) \in \mathfrak{p}'$. ⊣

COROLLARY 8.4. *For any stable formula $\varphi(x, y) \in L$, the following are equivalent.*

1. *$\varphi(x, a)$ does not divide over A.*
2. *$\varphi(x, a)$ is satisfiable in every model $M \supseteq A$.*
3. *Some $\mathfrak{p} \in S_\varphi(\mathfrak{C})$ containing $\varphi(x, a)$ is definable over* $\mathrm{acl}^{\mathrm{eq}}(A)$.
4. *$\varphi(x, a)$ does not fork over any model $M \supseteq A$.*
5. *Some $\mathfrak{p} \in S_\varphi(\mathfrak{C})$ containing $\varphi(x, a)$ does not divide over A.*

PROOF. The equivalence of 1, 2 and 3 follows directly from Proposition 8.3 applied to the empty type $\pi(x)$.

$2 \Rightarrow 4$ is a consequence of point 2 of Remark 4.4.

$4 \Rightarrow 1$ follows from point 7 of Remark 4.2.

$5 \Rightarrow 1$ is clear.

$3 \Rightarrow 5$. Assume $\mathfrak{p} \in S_\varphi(\mathfrak{C})$ is as in 3. We check that $\mathfrak{p}$ does not divide over A. Let $\varphi(x, a_1), \ldots, \varphi(x, a_n), \neg\varphi(x, b_1), \ldots, \neg\varphi(x, b_k) \in \mathfrak{p}$ and let $\psi(x; y_1, \ldots, y_n, z_1, \ldots, z_k)$ be the formula

$$\varphi(x, y_1) \wedge \cdots \wedge \varphi(x, y_n) \wedge \neg\varphi(x, z_1) \wedge \cdots \wedge \neg\varphi(x, z_k).$$

Notice that $\psi(x; y_1, \ldots, y_n, z_1, \ldots, z_k)$ is stable. Let $\mathfrak{q}(x) = \{\psi(x, c) : \mathfrak{p}(x) \vdash \psi(x, c)\} \cup \{\neg\psi(x, c) : \mathfrak{p}(x) \vdash \neg\psi(x, c)\}$. Then $\mathfrak{q} \in S_\psi(\mathfrak{C})$ is definable over $\mathrm{acl}^{\mathrm{eq}}(A)$ and $\psi(x; a_1, \ldots, a_n, b_1, \ldots, b_k) \in \mathfrak{q}$. By the equivalence $1 \Leftrightarrow 3$ applied to $\psi(x; a_1, \ldots, a_n, b_1, \ldots, b_k)$, $\psi(x; a_1, \ldots, a_n, b_1, \ldots, b_k)$ does not divide over A. Hence $\mathfrak{p}$ does not divide over A. ⊣

PROPOSITION 8.5. *Let T be simple and let $\Delta = \{\varphi_i(x, y_i) : i < n\}$ be a set of stable formulas. The following are equivalent for any Δ-type $\pi(x)$.*

1. *$\pi(x)$ does not fork over A.*
2. *$\pi(x)$ is finitely satisfiable in every model $M \supseteq A$.*
3. *Some $\mathfrak{p}(x) \in S_\Delta(\mathfrak{C})$ containing $\pi(x)$ is definable over* $\mathrm{acl}^{\mathrm{eq}}(A)$.

PROOF. Recall that if each $\varphi_i(x, y_i)$ is stable, then any boolean combination $\psi(x; y_1, \ldots, y_n)$ of the formulas $\varphi_i(x, y_i)$ is stable. Since dividing and forking are the same in a simple theory, $1 \Leftrightarrow 2$ follows from Corollary 8.4.

$1 \Rightarrow 3$. If $\pi(x)$ does not fork over A, it can be extended to some $\mathfrak{p}(x) \in S_\Delta(\mathfrak{C})$ that does not fork over A. By $1 \Leftrightarrow 2$, $\mathfrak{p}(x)$ is finitely satisfiable in every model $M \supseteq A$ and by Proposition 8.1 it is definable over every model $M \supseteq A$. This clearly implies that it is definable over $\mathrm{acl}^{\mathrm{eq}}(A)$.

$3 \Rightarrow 2$. By Proposition 8.1, $\mathfrak{p}$ is finitely satisfiable in every $M \supseteq A$ and hence the same is true for $\pi(x)$. ⊣

COROLLARY 8.6. *Let T be stable, $p(x) \in S(A)$ and $M \subseteq A$ (possibly $A = \mathfrak{C}$). The following are equivalent:*

1. *$p(x)$ does not fork over M.*
2. *$p(x)$ inherits from M.*
3. *$p(x)$ coinherits from M.*
4. *$p(x)$ is M-definable.*
5. *Some $\mathfrak{p} \in S(\mathfrak{C})$ extending p does not split over M.*

PROOF. The equivalence between points 2, 3, and 4 follows from Corollary 7.14. The equivalence of 4 and 5 follows from Proposition 7.12. The equivalence of 1 and 3 follows from Corollary 8.4. ⊣

COROLLARY 8.7. *Let T be stable and let $\mathfrak{p}(x) \in S(\mathfrak{C})$. The following are equivalent:*

1. *$\mathfrak{p}(x)$ does not fork over A.*
2. *$\mathfrak{p}(x)$ coinherits from every $M \supseteq A$.*
3. *$\mathfrak{p}(x)$ inherits from every $M \supseteq A$.*
4. *$\mathfrak{p}(x)$ does not split over any $M \supseteq A$.*
5. *$\mathfrak{p}(x)$ is definable over every $M \supseteq A$.*
6. *$\mathfrak{p}(x)$ is definable over $\mathrm{acl}^{\mathrm{eq}}(A)$.*
7. *The Cantor–Bendixson rank of $\mathfrak{p} \restriction \varphi$ in $S_\varphi(\mathfrak{C})$ is $\mathrm{CB}_\varphi(\mathfrak{p} \restriction A)$ for all φ.*
8. *$\mathfrak{p}(x)$ has a bounded orbit in $\mathrm{Aut}(\mathfrak{C}/A)$ (in fact of size $\leq 2^{|T|}$ if it is finitary).*

PROOF. The equivalence between points 2, 3, 4, 5, and 6 follows from Corollary 8.6 (for 6 observe that $\mathfrak{p}$ is definable over $\mathrm{acl}^{\mathrm{eq}}(A)$ if and only if it is definable over every model $M \supseteq A$).

$1 \Leftrightarrow 6$ follows from Corollary 8.4.

$6 \Leftrightarrow 7$ follows from Corollary 6.15.

$7 \Rightarrow 8$. The orbit of $\mathfrak{p} \restriction \varphi$ is bounded by $\mathrm{Mlt}_\varphi(\mathfrak{p} \restriction A) < \omega$ and hence the orbit of $\mathfrak{p}$ is bounded by $2^{|T|}$.

$8 \Rightarrow 6$. Let c_φ be the canonical parameter of the definition of $\mathfrak{p} \restriction \varphi$. Since c_φ has bounded orbit in $\mathrm{Aut}(\mathfrak{C}/A)$, in fact it has finite orbit. Hence $c_\varphi \in \mathrm{acl}^{\mathrm{eq}}(A)$ and $\mathfrak{p} \restriction \varphi$ is definable over $\mathrm{acl}^{\mathrm{eq}}(A)$. ⊣

COROLLARY 8.8. *Let T be stable, $p(x) \in S(A)$ and $\varphi(x, y) \in L$. The following are equivalent:*

1. *$p(x) \cup \{\varphi(x, a)\}$ does not fork over A.*
2. *$\mathrm{CB}_\psi(p(x) \cup \{\varphi(x, a)\}) = \mathrm{CB}_\psi(p)$ for all ψ.*
3. *$\mathrm{CB}_\varphi(p(x) \cup \{\varphi(x, a)\}) = \mathrm{CB}_\varphi(p)$.*

PROOF. $1 \Rightarrow 2$. Let $\mathfrak{p}(x) \in S(\mathfrak{C})$ be an extension of $p(x) \cup \{\varphi(x,a)\}$ which does not fork over A. By Corollary 8.7, $\mathrm{CB}_\psi(p(x))$ is the Cantor–Bendixson rank of $\mathfrak{p}(x) \restriction \psi$ in $S_\psi(\mathfrak{C})$. Hence $\mathrm{CB}_\psi(p(x) \cup \{\varphi(x,a)\}) = \mathrm{CB}_\psi(p(x))$.

$2 \Rightarrow 3$ is obvious. We prove $3 \Rightarrow 1$. Let $\mathfrak{p}(x) \in S(\mathfrak{C})$ be a nonforking extension of $p(x)$. By Corollary 8.7, $\mathrm{CB}_\varphi(p(x))$ is the Cantor–Bendixson rank of $\mathfrak{p}(x) \restriction \varphi$. Let $\mathfrak{q}(x) \in S(\mathfrak{C})$ be such that $\mathfrak{q}(x) \restriction \varphi$ is consistent with $p(x) \cup \{\varphi(x,a)\}$ and has Cantor–Bendixson rank $\mathrm{CB}_\varphi(p(x) \cup \{\varphi(x,a)\})$. By corollaries 6.14 and 6.15 $\mathfrak{p}(x) \restriction \varphi$ and $\mathfrak{q}(x) \restriction \varphi$ are A-conjugate. Since $\varphi(x,a) \in \mathfrak{q}(x) \restriction \varphi$, $p(x) \cup \{\varphi(x,a)\}$ is contained in an A-conjugate of $\mathfrak{p}(x)$, a global type which does not fork over A. Hence $p(x) \cup \{\varphi(x,a)\}$ does not fork over A. ⊣

COROLLARY 8.9. *Let T be stable, $A \subseteq B$ and $p(x) \in S(B)$. The following are equivalent*:

1. *$p(x)$ does not fork over A.*
2. *$\mathrm{CB}_\varphi(p) = \mathrm{CB}_\varphi(p \restriction A)$ for all φ.*

PROOF. It is an immediate consequence of Corollary 8.8. ⊣

PROPOSITION 8.10. *Let T be simple. If $\varphi(x,y) \in L$ is stable, for every A, a there is some $\sigma(x) \in L(A)$ equivalent to a positive boolean combination of A-conjugates of $\varphi(x,a)$ and such that for every $p(x) \in S(A)$, $\sigma(x) \in p(x)$ if and only if $p(x) \cup \{\varphi(x,a)\}$ does not fork over A.*

PROOF. Apply Proposition 8.3 with $p(x) = \pi(x)$. ⊣

COROLLARY 8.11 (Open mapping theorem). *Let T be stable and let $A \subseteq B$. The set $\mathrm{NF}(B,A)$ of all $p(x) \in S(B)$ which do not fork over A is closed in $S(B)$ and the restriction mapping $p \mapsto p \restriction A$ from $\mathrm{NF}(B,A)$ onto $S(A)$ is open.*

PROOF. The restriction map from $S(\mathfrak{C})$ onto $S(B)$ is continuous and hence closed and the image of $\mathrm{NF}(\mathfrak{C},A)$ is $\mathrm{NF}(B,A)$. Hence it is enough to check that $\mathrm{NF}(\mathfrak{C},A)$ is closed. Now,

$$\mathrm{NF}(\mathfrak{C},A) = \bigcap_{M \supseteq A} \{\mathfrak{p} \in S(\mathfrak{C}) : \mathfrak{p} \text{ coinherits from } M\}$$

and for each M, $\{\mathfrak{p} \in S(\mathfrak{C}) : \mathfrak{p} \text{ coinherits from } M\}$ is closed since it is the closure of $\{\mathrm{tp}(a/\mathfrak{C}) : a \in M\}$. The fact that the restriction map from $\mathrm{NF}(B,A)$ onto $S(A)$ is open is an immediate consequence of Proposition 8.10. ⊣

COROLLARY 8.12. *If T is stable, any nonforking extension of a nonisolated type is nonisolated.*

PROOF. By Proposition 8.10 or Corollary 8.11. ⊣

COROLLARY 8.13. *If T is simple, any nonforking extension of a type which is not isolated by stable formulas is also not isolated by stable formulas.*

PROOF. By Proposition 8.10. ⊣

Chapter 9

LASCAR STRONG TYPES

Here we will consider relations R, meaning binary relations between tuples of $\mathfrak{C}$ of length α for some ordinal α. The ordinal α usually is intended to be a natural number, but we do not impose restrictions.

DEFINITION 9.1. A relation R is *bounded* if for some cardinal κ there is no sequence $(a_i : i < \kappa)$ such that $\neg R(a_i, a_j)$ for all $i < j < \kappa$. The relation is *finite* if this bound κ is in fact a natural number. Observe that for definable relations finiteness is equivalent to boundedness. Note also that bounded relations are always reflexive.

REMARK 9.2. *For every cardinal λ, any intersection of λ bounded relations is a bounded relation.*

PROOF. Let $(R_l : l < \lambda)$ be a sequence of bounded relations. For all $l < \lambda$ let κ_l be a bound for R_l and let $\kappa = \lambda + \sup\{\kappa_l : l < \lambda\}$. Assume there are $(a_i : i < (2^\kappa)^+)$ such that $\neg R(a_i, a_j)$ for all $i < j < (2^\kappa)^+$, where $R = \bigcap_{l<\lambda} R_l$. By Erdős–Rado $((2^\kappa)^+ \to (\kappa^+)^2_\kappa)$ for some $l < \lambda$ there is a subset $I \subseteq (2^\kappa)^+$ of cardinality κ^+ such that $\neg R_l(a_i, a_j)$ for all $i < j$ in I. This contradicts the choice of κ_l. ⊣

LEMMA 9.3. 1. *The number of A-invariant relations on tuples of length α is at most $2^{2^{|T|+|A|+|\alpha|}}$.*

2. *There is a smallest A-invariant bounded equivalence relation (among tuples of a fixed length).*

PROOF. 1. By Lemma 1.2 the number of sets of types over A of tuples of length α is an upper bound. 2 follows from Remark 9.2. ⊣

DEFINITION 9.4. We say that the tuples a, b have the same *Lascar strong type over A* and we write $a \overset{\mathrm{Ls}}{\equiv}_A b$ if a and b are equivalent in the least A-invariant bounded equivalence relation. The *Lascar strong type of a over A* can be defined as the equivalence class $\mathrm{Lstp}(a/A)$ of a in $\overset{\mathrm{Ls}}{\equiv}_A$. Hence

$$a \overset{\mathrm{Ls}}{\equiv}_A b \Leftrightarrow \mathrm{Lstp}(a/A) = \mathrm{Lstp}(b/A).$$

In the case $A = \emptyset$ we omit the subscript.

DEFINITION 9.5. Let x, y be finite tuples of variables of the same length. We say that the formula $\theta(x, y)$ is *thick* if it defines a relation which is finite and symmetric. Note that all thick formulas define reflexive relations. For any set A and for any tuples of variables x, y of the same length, the set of all thick formulas over A in (finite subtuples of) the variables x, y will be

$$\mathrm{nc}_A(x, y).$$

In the case $A = \emptyset$ we omit the subscript. Let $\mathrm{nc}^n(x, y)$ be the n-times composition of $\mathrm{nc}(x, y)$; more formally: for $n \geq 2$, $\mathrm{nc}^n_A(x, y)$ is the type

$$\exists y_1 \ldots y_{n-1}(\mathrm{nc}_A(x, y_1) \wedge \mathrm{nc}_A(y_1, y_2) \wedge \cdots \wedge \mathrm{nc}_A(y_{n-1}, y))$$

and we complete the definition taking $\mathrm{nc}^1_A(x, y) = \mathrm{nc}_A(x, y)$ and taking as $\mathrm{nc}^0_A(x, y)$ the equation $x = y$.

REMARK 9.6. 1. *The conjunction and the disjunction of thick formulas are thick formulas.*
2. *Any consequence of a thick formula is a finite formula.*
3. *If $\varphi(x, y)$ is finite, then $\varphi(x, y) \wedge \varphi(y, x)$ is thick.*

PROOF. It is an easy exercise. ⊣

LEMMA 9.7. *For any a, b, $\models \mathrm{nc}_A(a, b)$ if and only if a, b start an infinite A-indiscernible sequence.*

PROOF. If a, b start an infinite A-indiscernible sequence, then $\models \theta(a, b)$ for every thick formula $\theta(x, y)$ over A. Now assume $\models \mathrm{nc}_A(a, b)$. Let $p(x, y) = \mathrm{tp}(ab/A)$. By Ramsey's Theorem and compactness, to prove that a, b start an infinite A-indiscernible sequence it is enough to check that there is an infinite sequence $(a_i : i < \omega)$ such that $\models p(a_i, a_j)$ for all $i < j < \omega$. For this we have to prove for any $\varphi \in p$, the consistency of $\{\varphi(x_i, x_j) : i < j < \omega\}$. If this set of formulas is inconsistent, then $\neg\varphi(x, y)$ is finite and therefore $(\neg\varphi(x, y) \wedge \neg\varphi(y, x)) \in \mathrm{nc}_A(x, y)$. Hence $\models \neg\varphi(a, b)$, a contradiction. ⊣

PROPOSITION 9.8. $\mathrm{nc}_A(x, a)$ *does not divide over A.*

PROOF. If $(a_i : i < \omega)$ is A-indiscernible and $a = a_0$, then by Lemma 9.7, a is a realization of $\bigcup_{i<\omega} \mathrm{nc}_A(x, a_i)$. ⊣

PROPOSITION 9.9. *The relation $\overset{\mathrm{Ls}}{\equiv}_A$ of equality of Lascar strong type over A is the transitive closure of the relation of starting an A-indiscernible sequence. Hence it is defined by the infinite disjunction $\bigvee_n \mathrm{nc}^n_A(x, y)$.*

PROOF. Since the relation of starting an infinite indiscernible sequence is defined by the type $\mathrm{nc}_A(x, y)$ consisting of finite formulas, it is bounded. Hence its transitive closure E is also bounded. Since E is a bounded A-invariant equivalence relation, $\overset{\mathrm{Ls}}{\equiv}_A \subseteq E$. For the other direction it suffices to show that if a, b start an infinite A-indiscernible sequence then $a \overset{\mathrm{Ls}}{\equiv}_A b$. Let κ be a strict upper bound for the number of $\overset{\mathrm{Ls}}{\equiv}_A$-classes. Choose an

A-indiscernible sequence of length κ starting with a, b. If $a \stackrel{\text{Ls}}{\not\equiv}_A b$ then by A-invariance $a' \stackrel{\text{Ls}}{\not\equiv}_A b'$ for any a', b' in the sequence, which contradicts the choice of κ. ⊣

DEFINITION 9.10. For a, b tuples of the same length, we put $d_A(a, b) = 0$ if $a = b$. If $a \neq b$ and there is some $n < \omega$ for which there are infinite A-indiscernible sequences $I_1, \dots, I_n$ and tuples $a_1, \dots, a_{n-1}$ such that $a = a_1$, $b = a_n$ and for each $i < n$, $a_i, a_{i+1} \in I_i$, then we define $d_A(a, b)$ as the least such n. If there is no such n we put $d_A(a, b) = \infty$.

The *diameter over A* of a class X of tuples of the same length is

$$\text{diam}_A(X) = \sup\{d_A(a, b) : a, b \in X\}.$$

REMARK 9.11. 1. *$d_A(a, b) \leq n$ if and only if $\models \text{nc}^n_A(a, b)$.*

2. $a \stackrel{\text{Ls}}{\equiv}_A b \Leftrightarrow d_A(a, b) < \infty$.

PROOF. Clear by definition and Proposition 9.9. ⊣

LEMMA 9.12. 1. *If $\models \text{nc}_A(a, b)$, then there is a model $M \supseteq A$ such that $a \equiv_M b$.*

2. *If $a \equiv_M b$ for some model $M \supseteq A$, then $\models \text{nc}^2_A(a, b)$.*

PROOF. 1. Fix an infinite A-indiscernible sequence I starting with a, b. By Corollary 1.7 I is indiscernible over some model $M \supseteq A$. Then $a \equiv_M b$.

2. Assume that $a \equiv_M b$ for some model $M \supseteq A$. We show that $\models \exists z(\theta(a, z) \wedge \theta(b, z))$ for every thick formula $\theta(x, y)$ over A. Let n be the maximal length of a sequence $a_1, \dots, a_n$ such that $\models \neg\theta(a_i, a_j)$ for all $i < j \leq n$. We can find such $a_1, \dots, a_n$ in M. For some $i \leq n$, $\models \theta(a, a_i)$. Since $a \equiv_M b$ also $\models \theta(b, a_i)$. ⊣

PROPOSITION 9.13. *Equality of Lascar strong types over A is the transitive closure of the relation of having the same type over a model containing A.*

PROOF. Clear by Proposition 9.9 and Lemma 9.12. ⊣

DEFINITION 9.14. The group $\text{Autf}(\mathfrak{C}/A)$ of *strong automorphisms over A* of the monster model $\mathfrak{C}$ is the subgroup of $\text{Aut}(\mathfrak{C}/A)$ generated by the automorphisms fixing a small submodel containing A:

$$\text{Autf}(\mathfrak{C}/A) = \langle \bigcup_{M \supseteq A} \text{Aut}(\mathfrak{C}/M) \rangle.$$

COROLLARY 9.15. *$a \stackrel{\text{Ls}}{\equiv}_A b$ if and only if $f(a) = b$ for some $f \in \text{Autf}(\mathfrak{C}/A)$.*

PROOF. It follows from Proposition 9.13. ⊣

COROLLARY 9.16. *If $a \stackrel{\text{Ls}}{\equiv}_A b$ then for any c there is some d such that $ac \stackrel{\text{Ls}}{\equiv}_A bd$.*

PROOF. Choose $f \in \text{Autf}(\mathfrak{C}/A)$ such that $f(a) = b$ and put $d = f(c)$. ⊣

DEFINITION 9.17. Let $\pi(x)$ be a partial type over A such that for each $a, b \models \pi$ there is some $f \in \mathrm{Aut}(\mathfrak{C})$ fixing setwise $\pi(\mathfrak{C})$ such that $f(a) = b$. Let U be a class consisting of tuples of the length of x. We say that U is *c-free over* π if for some $n < \omega$ there are automorphisms $f_1, \ldots, f_n$ fixing setwise $\pi(\mathfrak{C})$ such that $\pi(\mathfrak{C}) \subseteq \bigcup_{i=1}^n f_i(U)$. U is called *weakly c-free over* π if for some V non c-free over π, the union UV is c-free over π. When π is fixed, we will sometimes omit to mention it. A formula is (weakly) c-free if it defines a (weakly) c-free class. A type is (weakly) c-free if it only implies (weakly) c-free formulas.

REMARK 9.18. *Let $\pi(x)$ be a partial type as in the definition of c-freeness.*

1. *If U is (weakly) c-free and $U \subseteq V$, V is also (weakly) c-free.*
2. *$\mathcal{I} = \{U : U$ is not weakly c-free$\}$ is an ideal (i.e. it is downward closed and closed under finite unions).*
3. *Any partial type over $B \supseteq A$ which is weakly c-free can be extended to a complete weakly c-free $p(x) \in S(B)$.*
4. *$\pi(x)$ is a c-free type.*

PROOF. For 3 note that every filter $\mathcal{F}$ such that $\mathcal{I} \cap \mathcal{F} = \emptyset$ can be extended to an ultrafilter $\mathcal{F}'$ such that $\mathcal{I} \cap \mathcal{F}' = \emptyset$. ⊣

LEMMA 9.19. *Let $\pi(x)$ be a partial type as in the definition of c-freeness. Let U be definable. If U is weakly c-free, then for some definable V, V is not c-free and UV is c-free.*

PROOF. To fix notation assume $U \subseteq \mathfrak{C}^\alpha$. There is some non c-free V such that UV is c-free and hence there are automorphisms $f_1, \ldots, f_n$ fixing setwise $\pi(\mathfrak{C})$ such that $\pi(\mathfrak{C}) \subseteq \bigcup_{i=1}^n f_i(UV)$. Let $W = \bigcap_{i=1}^n \mathfrak{C}^\alpha \smallsetminus f_i(U)$. It is definable. If f_{n+1} is the identity, then

$$\pi(\mathfrak{C}) \subseteq W \cup (\mathfrak{C}^\alpha \smallsetminus W) = W \cup \bigcup_{i=1}^n f_i(U) \subseteq UW \cup \bigcup_{i=1}^n f_i(UW) = \bigcup_{i=1}^{n+1} f_i(UW)$$

and therefore UW is c-free. Since $\bigcup_{i=1}^n f_i(V)$ is not c-free and $W \cap \pi(\mathfrak{C}) \subseteq \bigcup_{i=1}^n f_i(V)$, $W \cap \pi(\mathfrak{C})$ is not c-free either. This implies that W is not c-free. ⊣

PROPOSITION 9.20. *Let $\pi(x)$ be a type over A as in the definition of c-freeness and let $\varphi(x, y) \in L$. Assume T is simple.*

1. *If $\varphi(x, a)$ is weakly c-free over π, then $\pi(x) \cup \{\varphi(x, a)\}$ does not fork over A.*
2. *If $\varphi(x, y)$ is stable, $\pi(x)$ is a complete type over A, and $\pi(x) \cup \{\varphi(x, a)\}$ does not fork over A, then $\varphi(x, a)$ is c-free over π.*

PROOF. 1. Let $\varphi(x, a)$ be weakly c-free (over π). By Lemma 9.19 for some $\psi(x, z) \in L$, for some b, $\psi(x, b)$ is not c-free and $\varphi(x, a) \vee \psi(x, b)$ is c-free. There are automorphisms $f_1, \ldots, f_k$ fixing setwise $\pi(\mathfrak{C})$ such that $\pi(x) \vdash \varphi'(x, a') \vee \psi'(x, b')$, where $\varphi'(x, a') = \bigvee_{i=1}^k \varphi(x, f_i(a))$ and $\psi'(x, b') = \bigvee_{i=1}^k \psi(x, f_i(b))$. Note that $\psi'(x, b')$ is not c-free. Assume $\pi(x) \cup \{\varphi(x, a)\}$

forks over A. Then $\pi(x) \cup \{\varphi'(x, a')\}$ forks (and divides) over A too. For some finite conjunction $\theta(x)$ of formulas of π, for some sequence $(a'_i : i < \omega)$ of realizations of $\mathrm{tp}(a'/A)$, for some $l < \omega$,

$$\{\theta(x) \wedge \varphi'(x, a'_i) : i < \omega\}$$

is l-inconsistent. Let $g_1, \ldots, g_l$ be automorphisms fixing A pointwise and sending a' to $a'_1, \ldots, a'_l$ respectively. Then $\pi(x) \vdash \bigvee_{i=1}^{l} \psi'(x, g_i(b'))$, which implies that $\psi'(x, b')$ is c-free, a contradiction.

2. Let $\pi(x) = p(x) \in S(A)$ and assume $p(x) \cup \{\varphi(x, a)\}$ does not fork over A. By Proposition 8.10 there is a formula $\sigma(x) \in p$ which is equivalent to a positive boolean combination of A-conjugates of $\varphi(x, a)$, say to $\bigvee_{i=1}^{k} \psi_i(x, a_i)$ where each $\psi_i(x, a_i)$ is a conjunction of A-conjugates of $\varphi(x, a)$. If we select an A-conjugate $\varphi(x, a'_i)$ of $\varphi(x, a)$ in each $\psi_i(x, a_i)$ we see that $p(x) \vdash \bigvee_{i=1}^{k} \varphi(x, a'_i)$, which shows that $\varphi(x, a)$ is c-free over p. ⊣

PROPOSITION 9.21. *Let $\pi(x)$ be a partial type over A as in the definition of c-freeness. Consider the sets*

1. $P = \{q(x, y) \in S(\emptyset) : q(x, y) \cup \pi(x) \cup \pi(y)$ *is consistent*$\}$,
2. $P_w = \{q(x, y) \in P : q(a, y)$ *is weakly c-free for some (all)* $a \models \pi\}$.

Then P and P_w are nonempty closed subsets of $S(\emptyset)$. Moreover, if $S \subseteq P_w$ is nonempty and relatively open, then for some $k < \omega$ there are $c_1, \ldots, c_k \models \pi$ such that for all $b \models \pi$ there is some $d \models \pi$ such that $\mathrm{tp}(bd) \in S$ *and* $\mathrm{tp}(c_i d) \in S$ *for some i.*

PROOF. With respect to the first statement, only the case of P_w needs some verification. Fix $a \models \pi$. Note that $\pi(y)$ is weakly c-free and therefore it can be extended to a complete weakly c-free type $q(a, y)$. Thus $P_w \neq \emptyset$. For closedness, note that $P_w = \{q(x, y) \in S(\emptyset) : q(x, y) \supseteq \Gamma(x, y)\}$ if $\Gamma(x, y) = \{\neg\varphi(x, y) \in L : \varphi(a, y)$ is not weakly c-free$\}$.

Recall from chapter 1 that for any set of formulas $\Sigma(x, y)$ over $\emptyset$, $[\Sigma]$ denotes the corresponding closed set in $S(\emptyset)$. Now choose $\varphi(x, y) \in L$ witnessing that S is not empty and it is relatively open in P_w. Making S smaller if necessary, we can assume that

$$S = P_w \cap [\varphi(x, y)].$$

Fix $c \models \pi$. Note that $\varphi(c, y)$ is weakly c-free. By Lemma 9.19, there is a non c-free $\psi(e, y)$ such that $\varphi(c, y) \vee \psi(e, y)$ is c-free. Hence for some automorphisms $f_1, \ldots, f_k$ fixing setwise $\pi(\mathfrak{C})$,

$$(*) \quad \pi(y) \vdash \bigvee_{i=1}^{k} \varphi(c_i, y) \vee \psi(e_i, y)$$

where $c_i = f_i(c)$ and $e_i = f_i(e)$. Now, since $S \subseteq S(\emptyset)$ is closed, $S = [\rho(x, y)]$ for some partial type $\rho(x, y)$ over $\emptyset$. Note that $\rho(c, \mathfrak{C}) = \bigcup_{q \in S} q(c, \mathfrak{C})$.

CLAIM 1. *$(\varphi(c, \mathfrak{C}) \cup \psi(e, \mathfrak{C})) \setminus \rho(c, \mathfrak{C})$ is not c-free.*

Assume the claim is true and our proposition is false for k, for $c_1, \dots, c_k$ and for some $b \models \pi$. We will reach a contradiction. For all $d \in \rho(b, \mathfrak{C})$, $d \notin \bigcup_{i=1}^{k} \rho(c_i, \mathfrak{C})$, and by $(*)$, $d \in \bigcup_{i=1}^{k} \varphi(c_i, \mathfrak{C}) \cup \psi(e_i, \mathfrak{C})$ and therefore

$$\rho(b, \mathfrak{C}) \subseteq \bigcup_{i=1}^{k} ((\varphi(c_i, \mathfrak{C}) \cup \psi(e_i, \mathfrak{C})) \setminus \rho(c_i, \mathfrak{C})).$$

Choose an automorphism f_{k+1} fixing setwise $\pi(\mathfrak{C})$ and such that $f_{k+1}(c) = b$. Let $c_{k+1} = f_{k+1}(c)$ and $e_{k+1} = f_{k+1}(e)$. Since $\varphi(c_{k+1}, \mathfrak{C}) \cup \psi(e_{k+1}, \mathfrak{C})$ clearly is a subset of $\rho(c_{k+1}, \mathfrak{C}) \cup ((\varphi(c_{k+1}, \mathfrak{C}) \cup \psi(e_{k+1}, \mathfrak{C})) \setminus \rho(c_{k+1}, \mathfrak{C}))$, it follows that

$$\varphi(c_{k+1}, \mathfrak{C}) \cup \psi(e_{k+1}, \mathfrak{C}) \subseteq \bigcup_{i=1}^{k+1} (\varphi(c_i, \mathfrak{C}) \cup \psi(e_i, \mathfrak{C})) \setminus \rho(c_i, \mathfrak{C}).$$

Since $\varphi(c, \mathfrak{C}) \cup \psi(e, \mathfrak{C})$ is c-free, $\bigcup_{i=1}^{k+1} (\varphi(c_i, \mathfrak{C}) \cup \psi(e_i, \mathfrak{C})) \setminus \rho(c_i, \mathfrak{C})$ must be c-free and this contradicts the claim.

PROOF OF CLAIM 1. Assume $(\varphi(c, \mathfrak{C}) \cup \psi(c, \mathfrak{C})) \setminus \rho(c, \mathfrak{C})$ is c-free. There are automorphisms $g_1, \dots, g_n$ fixing setwise $\pi(\mathfrak{C})$ such that (with notation $c'_i = g_i(c)$ and $e'_i = g_i(e)$),

$$\pi(\mathfrak{C}) \subseteq \bigcup_{i=1}^{n} (\varphi(c'_i, \mathfrak{C}) \cup \psi(e'_i, \mathfrak{C})) \setminus \rho(c'_i, \mathfrak{C})$$

that is, $\pi(y) \wedge \bigwedge_{i=1}^{n} ((\neg\varphi(c'_i, y) \wedge \neg\psi(e'_i, y)) \vee \rho(c'_i, y))$ is inconsistent. By compactness, there are $\alpha_i(x, y) \in \rho(x, y)$ for $i = 1, \dots, n$ such that $\alpha_i(x, y) \vdash \varphi(x, y)$ and $\pi(y) \wedge \bigwedge_{i=1}^{n} ((\neg\varphi(c'_i, y) \wedge \neg\psi(e'_i, y)) \vee \alpha_i(c'_i, y))$ is inconsistent. Let $\alpha(x, y) = \bigwedge_{i=1}^{n} \alpha_i(x, y)$. Then also

$$\pi(y) \wedge \bigwedge_{i=1}^{n} ((\neg\varphi(c'_i, y) \wedge \neg\psi(e'_i, y)) \vee \alpha(c'_i, y))$$

is inconsistent, that is $\pi(\mathfrak{C}) \subseteq \bigcup_{i=1}^{n} (\varphi(c'_i, \mathfrak{C}) \cup \psi(e'_i, \mathfrak{C})) \setminus \alpha(c'_i, \mathfrak{C})$, which implies that $(\varphi(c, y) \vee \psi(e, y)) \wedge \neg\alpha(c, y)$ is c-free. Then also $(\varphi(c, y) \wedge \neg\alpha(c, y)) \vee \psi(e, y)$ is c-free and $(\varphi(c, y) \wedge \neg\alpha(c, y))$ is weakly c-free and can be extended to some weakly c-free complete type $q(c, y)$. This is a contradiction, because $q(x, y) \in P_w$ and $\varphi(x, y) \in q(x, y)$, which implies $q(x, y) \in P_w \cap [\varphi(x, y)] = S = [\rho(x, y)] \subseteq [\alpha(x, y)]$. ⊣

⊣

THEOREM 9.22. *A type-definable Lascar strong type has finite diameter.*

PROOF. Let $\pi(x)$ be a partial type defining $\mathrm{Lstp}(a/A)$ over aA. Then $\pi(x)$ satisfies our assumptions in the definition of c-freeness. Let P, P_w be as in Proposition 9.21. Let $X_n = P \cap [\mathrm{nc}^n(x, y)]$. Then, by Proposition 9.9, $P = \bigcup_{n \in \omega} X_n$. Since P_w is a nonempty closed set, by the Baire Category

Theorem, some X_n has nonempty interior in P_w. By the Proposition 9.21, there is some $k < \omega$ and $c_1, \ldots, c_k \models \pi$ such that for all $a, b \models \pi$ there are $c, d \models \pi$ such that $\mathrm{tp}(a,c), \mathrm{tp}(b,d) \in X_n$ and $\mathrm{tp}(c_i, c), \mathrm{tp}(c_j, d) \in X_n$ for some $i, j \leq k$. Let $m = \max\{d(c_l, c_{l'}) : 1 \leq l, l' \leq k\}$. Then

$$d(a,b) \leq d(a,c) + d(c,c_i) + d(c_i,c_j) + d(c_j,d) + d(d,b) \leq 4n + m. \quad \dashv$$

COROLLARY 9.23. 1. *If for each $n < \omega$ there is a Lascar strong type over A of diameter $\geq n$, then there is a Lascar strong type over A which is not type-definable.*
2. *Equality of Lascar strong types over A is type-definable if and only if for some $n < \omega$ it is defined by* $\mathrm{nc}_A^n(x,y)$.

PROOF. 1. For each $n < \omega$ fix a tuple a_n whose Lascar strong type over A has diameter at least n and consider $a = (a_n : n < \omega)$. It is easy to check that $\mathrm{Lstp}(a/A)$ has infinite diameter. By Theorem 9.22, it is not type-definable.

2. Assume the relation $\overset{\mathrm{Ls}}{\equiv}_A$ of equality of Lascar strong types is type-definable. By 1 there is a bound $n < \omega$ for the diameter of any Lascar strong type over A. Since $d_A(a,b) \leq n$ is equivalent to $\models \mathrm{nc}_A^n(a,b)$, the type $\mathrm{nc}_A^n(x,y)$ defines $\overset{\mathrm{Ls}}{\equiv}_A$. $\dashv$

DEFINITION 9.24. As in the case of A-invariance, there is a smallest type-definable over A bounded equivalence relation (on tuples of a given length). We say that the tuples a, b have the same *Kim–Pillay strong type over A* or the same *bounded type over A* and we write $a \overset{\mathrm{KP}}{\equiv}_A b$ if a and b are equivalent in the least type-definable over A bounded equivalence relation.

We say that a, b have the same *strong type over A* and we write $a \overset{\mathrm{s}}{\equiv}_A b$ if a and b are equivalent in every A-definable finite equivalence relation. As usual, in case $A = \emptyset$ we omit the subscript.

REMARK 9.25. 1. *If $a \overset{\mathrm{Ls}}{\equiv}_A b$, then $a \overset{\mathrm{KP}}{\equiv}_A b$.*
2. *If $a \overset{\mathrm{KP}}{\equiv}_A b$, then $a \overset{\mathrm{s}}{\equiv}_A b$.*
3. *If $a \overset{\mathrm{s}}{\equiv}_A b$, then $a \equiv_A b$.*

PROOF. 1 is clear since every equivalence relation that is type-definable over A is A-invariant. Similarly for 2, since every A-definable finite equivalence relation is bounded and type-definable over A. For 3 observe that for each $\varphi(x) \in L(A)$, the equivalence relation $E(x,y)$ defined by $(\varphi(x) \leftrightarrow \varphi(y))$ is A-definable and has only two classes. $\dashv$

DEFINITION 9.26. The *strong type* of a over A is defined by

$$\mathrm{stp}(a/A) = \mathrm{tp}(a/\mathrm{acl}^{\mathrm{eq}}(A)).$$

LEMMA 9.27. $\mathrm{stp}(a/A) = \mathrm{stp}(b/A)$ *if and only if* $a \overset{\mathrm{s}}{\equiv}_A b$.

PROOF. Assume $\text{stp}(a/A) = \text{stp}(b/A)$. Let E be a finite A-definable equivalence relation, say defined by $\varphi(x, y, c)$ where $c \in A$ and $\varphi(x, y, z) \in L$. Let $\psi(z) \in \text{tp}(c)$ be a formula expressing that $\varphi(x, y, z)$ defines an equivalence relation in x, y and consider the relation $F(ux; vy)$ defined by

$$F(ux; vy) \Leftrightarrow (\neg\psi(u) \wedge \neg\psi(v)) \vee (\psi(u) \wedge u = v \wedge \varphi(x, y, u)).$$

It is a 0-definable equivalence relation and therefore the equivalence classes $[ca]_F$ and $[cb]_F$ are imaginary elements. Since $F(cx; cy)$ defines E and E is finite, these imaginaries are algebraic over A, that is, they are elements of $\text{acl}^{\text{eq}}(A)$. This clearly implies $[ca]_F = [cb]_F$ and therefore $E(a, b)$.

For the other direction, notice that according to Proposition 1.11 a relation R defined by a formula $\varphi(x) \in \text{acl}^{\text{eq}}(A)$ has finitely many A-conjugates and it is therefore a union of classes of some finite A-definable equivalence relation. $\dashv$

PROPOSITION 9.28. *Let T be stable. If $a \overset{\text{s}}{\equiv}_A b$, $A \subseteq B$, $a \mathop{\smile\!\!\!\!\!\!\mid}_A B$ and $b \mathop{\smile\!\!\!\!\!\!\mid}_A B$, then $a \overset{\text{s}}{\equiv}_B b$.*

PROOF. Let $p(x) = \text{stp}(a/A) = \text{stp}(b/A)$, let $\mathfrak{p} \in S(\mathfrak{C})$ be a nonforking extension of $\text{stp}(a/B)$ and let $\mathfrak{q} \in S(\mathfrak{C})$ be a nonforking extension of $\text{stp}(b/B)$. Since $a \mathop{\smile\!\!\!\!\!\!\mid}_A B$, $a \mathop{\smile\!\!\!\!\!\!\mid}_A \text{acl}^{\text{eq}}(B)$ and therefore $\mathfrak{p}$ does not fork over A. By Corollary 8.7 $\mathfrak{p}$ is definable over $\text{acl}^{\text{eq}}(A)$. By the same argument $\mathfrak{q}$ is definable over $\text{acl}^{\text{eq}}(A)$ and by Proposition 6.13 $\mathfrak{p} = \mathfrak{q}$. Hence $\text{stp}(a/B) = \text{stp}(b/B)$. $\dashv$

COROLLARY 9.29. *If T is stable, then $\overset{\text{Ls}}{\equiv}_A = \overset{\text{s}}{\equiv}_A$ for every A.*

PROOF. Let $a \overset{\text{s}}{\equiv}_A b$. Choose $M \supseteq A$ such that $M \mathop{\smile\!\!\!\!\!\!\mid}_A ab$. Then $a \mathop{\smile\!\!\!\!\!\!\mid}_A M$ and $b \mathop{\smile\!\!\!\!\!\!\mid}_A M$. By Proposition 9.28 $a \equiv_M b$ and hence by Lemma 9.12 and Proposition 9.9, $a \overset{\text{Ls}}{\equiv}_A b$. $\dashv$

THEOREM 9.30 (Finite equivalence relation theorem). *Let T be a stable theory. Let $A \subseteq B$, $r(x) \in S(A)$, and let $p(x), q(x) \in S(B)$ be two different nonforking extensions of r. Then for some $\varphi(x) \in L(B)$ equivalent to a formula over $\text{acl}^{\text{eq}}(A)$, $\varphi \in p$ while $\neg\varphi \in q$. There is also a finite A-definable equivalence relation E such that*

$$p(x) \cup q(y) \vdash \neg E(x, y).$$

PROOF. Let $p'(x) \in S(B \cup \text{acl}^{\text{eq}}(A))$ be an extension of p. If $p'(x) \upharpoonright \text{acl}^{\text{eq}}(A) \cup q(x)$ is consistent then there is some extension $q'(x) \in S(B \cup \text{acl}^{\text{eq}}(A))$ of q such that $p' \upharpoonright \text{acl}^{\text{eq}}(A) = q' \upharpoonright \text{acl}^{\text{eq}}(A)$. But then p' and q' are different nonforking extensions of the same strong type, which contradicts Corollary 9.29. Hence $p'(x) \upharpoonright \text{acl}^{\text{eq}}(A) \cup q(x)$ is inconsistent and there is some $\psi(x) \in p'(x) \upharpoonright \text{acl}^{\text{eq}}(A)$ such that $q(x) \vdash \neg\psi(x)$. Let $\varphi(x)$ be the disjunction of all B-conjugates of ψ. Then $p(x) \vdash \varphi(x)$, $q(x) \vdash \neg\varphi(x)$ and $\varphi(x) \in L(\text{acl}^{\text{eq}}(A))$ is equivalent to a formula over B.

With respcct to the last assertion, by Proposition 1.11 $\varphi(x)$ defines a union of classes of a finite A-definable equivalence relation E, and then clearly $p(x) \cup q(y) \vdash \neg E(x, y)$. $\dashv$

Chapter 10

THE INDEPENDENCE THEOREM

LEMMA 10.1. *Let T be simple and let $\kappa > |T|$. If $(a_i : i < \kappa)$ is A-independent, then for any set B of cardinality $< \kappa$ there is some $i < \kappa$ such that $B \mathop{\smile\!\!\!\!\!\!|}_A a_i$.*

PROOF. By choice of κ, there is a proper subset $X \subseteq \kappa$ such that $B \mathop{\smile\!\!\!\!\!\!|}_{Aa_X} \{a_i : i < \kappa\}$ where $a_X = (a_i : i \in X)$. Take $i \notin X$. Then $B \mathop{\smile\!\!\!\!\!\!|}_{Aa_X} a_i$ and, by Corollary 5.21, $a_i \mathop{\smile\!\!\!\!\!\!|}_A a_X$. By symmetry and transitivity, $B \mathop{\smile\!\!\!\!\!\!|}_A a_i$. ⊣

LEMMA 10.2. *Let T be simple. For any a, A and $B \supseteq A$ there is some a' such that $a' \overset{\text{Ls}}{\equiv}_A a$ and $a' \mathop{\smile\!\!\!\!\!\!|}_A B$.*

PROOF. Let κ be a cardinal larger than $|T| + |B|$. Let $(a_i : i < \kappa)$ be a Morley sequence in $\text{tp}(a/A)$ starting with $a_0 = a$. By Lemma 10.1 there is some $i < \kappa$ such that $B \mathop{\smile\!\!\!\!\!\!|}_A a_i$. Clearly, $a \overset{\text{Ls}}{\equiv}_A a_i$. ⊣

LEMMA 10.3. *Let T be simple and let $a \overset{\text{Ls}}{\equiv}_A b$. For any c, B there is some d such that $ac \overset{\text{Ls}}{\equiv}_A bd$ and $d \mathop{\smile\!\!\!\!\!\!|}_{Ab} B$.*

PROOF. By Corollary 9.16 there is some d' such that $ac \overset{\text{Ls}}{\equiv}_A bd'$ and by Corollary 9.15, there is a strong automorphism $f \in \text{Autf}(\mathfrak{C}/A)$ such that $f(ac) = bd'$. By Lemma 10.2 there is some d such that $d \overset{\text{Ls}}{\equiv}_{Ab} d'$ and $d \mathop{\smile\!\!\!\!\!\!|}_{Ab} B$. Again by Corollary 9.15 there is some $g \in \text{Autf}(\mathfrak{C}/Ab)$ such that $g(d') = d$. It follows that $g \circ f \in \text{Autf}(\mathfrak{C}/A)$ and $g \circ f(ac) = bd$. Hence $ac \overset{\text{Ls}}{\equiv}_A bd$. ⊣

LEMMA 10.4. *Let T be simple. If $(a_i : i < \omega+\omega)$ is an infinite A-indiscernible sequence, then $(a_i : \omega \leq i < \omega + \omega)$ is a Morley sequence over $A\{a_i : i < \omega\}$.*

PROOF. Let $a = (a_i : i < \omega)$. Clearly $(a_i : \omega \leq i < \omega + \omega)$ is Aa-indiscernible. It suffices to show that it is Aa-independent. Let X be a finite subset of $\{i : \omega \leq i < \omega + \omega\}$ an let $i < \omega + \omega$ be larger than every element in X. By symmetry it will be enough to check that $a_X \mathop{\smile\!\!\!\!\!\!|}_{Aa} a_i$, where $a_X = (a_j : j \in X)$. But this is clear since by A-indiscernibility $\text{tp}(a_X/Aaa_i)$ is finitely satisfiable in a. ⊣

PROPOSITION 10.5. *Let T be simple and let $\pi(x, y)$ be a set of formulas over A. If $(a_i : i \in I)$ is an A-indiscernible sequence and $\pi(x, a_i)$ does not fork over A for some (every) $i \in I$, then $\bigcup_{i\in I} \pi(x, a_i)$ does not fork over A.*

PROOF. We may assume $I = \omega$ and $\pi(x, a_0)$ does not fork over A. Let us first assume that $(a_i : i < \omega)$ is a Morley sequence over A. Since $\pi(x, a_0)$ does not divide over A, $\bigcup_{i<\omega} \pi(x, a_i)$ is consistent. Let $n < \omega$ and let $\Phi(x, y_0, \dots, y_{n-1}) = \pi(x, y_1) \cup \dots \cup \pi(x, y_n)$. We will show that $\Phi(x, a_0, \dots, a_{n-1})$ does not divide over A. If $b_i = a_{n\cdot i} \dots a_{n\cdot i+n-1}$, then $(b_i : i < \omega)$ is an infinite Morley sequence in $\mathrm{tp}(b_0/A)$ and $\bigcup_{i<\omega} \Phi(x, b_i)$ is consistent. By Proposition 5.15, $\Phi(x, b_0)$ does not divide over A.

Now let us consider the general case, where $(a_i : i < \omega)$ is just an A-indiscernible sequence. Choose $b = (b_i : i < \omega)$ such that $(b_i : i < \omega)^\frown(a_i : i < \omega)$ is A-indiscernible. By Lemma 10.4 $(a_i : i < \omega)$ is a Morley sequence over $A \cup b$. Let $p(x, y) \in S(Ab)$ be such that $p(x, a_0)$ extends $\pi(x, a_0)$ and does not fork over A. Then it does not fork over Ab and by the first case, $\bigcup_{i<\omega} p(x, a_i)$ does not fork over Ab. Let $c \models \bigcup_{i<\omega} p(x, a_i)$ be such that $c \mathop{\smile\!\!\!\!\!\!|}_{Ab} (a_i : i < \omega)$. Since $p(x, a_0)$ does not fork over A, also $c \mathop{\smile\!\!\!\!\!\!|}_A ba_0$. Hence $c \mathop{\smile\!\!\!\!\!\!|}_A b(a_i : i < \omega)$, which shows that $\bigcup_{i<\omega} \pi(x, a_i)$ does not fork over A. $\dashv$

LEMMA 10.6. *Assume a, b start an infinite A-indiscernible sequence and $c \mathop{\smile\!\!\!\!\!\!|}_{Aa} b$. Then for some d, the extended tuples ac, bd also start an infinite A-indiscernible sequence.*

PROOF. We may assume $A = \emptyset$. Let $c \mathop{\smile\!\!\!\!\!\!|}_a b$ and assume $(a_i : i < \omega)$ is an infinite indiscernible sequence with $a = a_0$ and $b = a_1$. Since $(a_n : n \geq 1)$ is a-indiscernible and $c \mathop{\smile\!\!\!\!\!\!|}_a b$, by Lemma 4.7 there is an ac-indiscernible sequence $(a'_n : n \geq 1)$ such that $(a_n : n \geq 1) \equiv_{ab} (a'_n : n \geq 1)$. Thus we may assume that $a_n = a'_n$ for all $n \geq 1$. Let $c_0 = c$ and choose for $n \geq 1$ some c_n such that

$$ca_0a_1 \dots \equiv c_n a_n a_{n+1} \dots .$$

Since $(a_n : n \geq 1)$ is ac-indiscernible, $cab \equiv caa_m$. Hence $cab \equiv c_n a_n a_{n+m}$, i.e., in the sequence $(c_n a_n : n < \omega)$ all triangles $c_n a_n a_{n+m}$ have the same type $p(x, y, z) = \mathrm{tp}(cab)$. By Ramsey's Theorem there is an indiscernible sequence $(d_n b_n : n < \omega)$ where all triangles $d_n b_n b_{n+m}$ satisfy $p(x, y, z)$. Clearly we may assume that $c = d_0$, $a = b_0$ and $b = b_1$. Take $d = d_1$. $\dashv$

PROPOSITION 10.7. *Let T be simple, let $\varphi(x, y), \psi(x, z) \in L(A)$, and assume that $\varphi(x, a) \wedge \psi(x, b)$ does not fork over A. If b, b' start an infinite A-indiscernible sequence and $a \mathop{\smile\!\!\!\!\!\!|}_{Ab} b'$, then $\varphi(x, a) \wedge \psi(x, b')$ does not fork over A.*

PROOF. Apply Lemma 10.6 finding a' such that $ba, b'a'$ start an infinite A-indiscernible sequence. By Proposition 10.5, $\varphi(x, a) \wedge \psi(x, b) \wedge \varphi(x, a') \wedge \psi(x, b')$ does not fork over A. In particular $\varphi(x, a) \wedge \psi(x, b')$ does not fork over A. $\dashv$

COROLLARY 10.8. *Let T be simple, let* $\varphi(x,y), \psi(x,z) \in L(A)$, *and assume that* $\varphi(x,a) \wedge \psi(x,b)$ *does not fork over* A. *If* $b \stackrel{\text{Ls}}{\equiv}_A b'$, $a \mathop{\smile\!\!\!\!\!\mid}_A b$, *and* $a \mathop{\smile\!\!\!\!\!\mid}_A b'$, *then* $\varphi(x,a) \wedge \psi(x,b')$ *does not fork over* A.

PROOF. Note that we can assume $a \mathop{\smile\!\!\!\!\!\mid}_A bb'$ since by Lemma 10.2 we can replace b' by some $b'' \mathop{\smile\!\!\!\!\!\mid}_{Aa} b$ such that $b'' \stackrel{\text{Ls}}{\equiv}_{Aa} b'$, which implies $b''b \mathop{\smile\!\!\!\!\!\mid}_A a$. Find $b_1, \dots, b_n$ such that $b = b_1$, $b' = b_n$ and b_i, b_{i+1} start an infinite A-indiscernible sequence. Let a' be such that $a' \equiv_{Abb'} a$ and $a' \mathop{\smile\!\!\!\!\!\mid}_{Abb'} b_1, \dots, b_n$. By Proposition 10.7 we see that $\varphi(x,a') \wedge \psi(x,b_i)$ does not fork over A for all $i \leq n$. Hence $\varphi(x,a) \wedge \psi(x,b')$ does not fork over A. ⊣

COROLLARY 10.9 (Independence Theorem). *Let T be a simple theory, let* $\varphi(x,y), \psi(x,z) \in L(A)$, *and assume* $a \mathop{\smile\!\!\!\!\!\mid}_A b$. *If there are* c, d *such that*

$$\models \varphi(c,a),\ c \mathop{\smile\!\!\!\!\!\mid}_A a,\ \models \psi(d,b),\ d \mathop{\smile\!\!\!\!\!\mid}_A b,\ \text{and } c \stackrel{\text{Ls}}{\equiv}_A d,$$

then $\varphi(x,a) \wedge \psi(x,b)$ *does not fork over* A.

PROOF. Using Lemma 10.3, choose $b' \mathop{\smile\!\!\!\!\!\mid}_{Ac} ab$ such that $cb' \stackrel{\text{Ls}}{\equiv}_A db$. Then $\models \varphi(c,a) \wedge \psi(c,b')$ and $c \mathop{\smile\!\!\!\!\!\mid}_A ab'$. Therefore $\varphi(x,a) \wedge \psi(x,b')$ does not fork over A. Since $a \mathop{\smile\!\!\!\!\!\mid}_A bb'$ by Corollary 10.8, $\varphi(x,a) \wedge \psi(x,b)$ does not fork over A. ⊣

COROLLARY 10.10. *Let T be simple.*

1. *Assume* A *is a common subset of* B *and* C. *Assume* $B \mathop{\smile\!\!\!\!\!\mid}_A C$, *and let* $b \mathop{\smile\!\!\!\!\!\mid}_A B$, $c \mathop{\smile\!\!\!\!\!\mid}_A C$, *be such that* $b \stackrel{\text{Ls}}{\equiv}_A c$. *Then for some* $d \mathop{\smile\!\!\!\!\!\mid}_A BC$, $d \equiv_B b$ *and* $d \equiv_C c$.
2. *Let* $(a_i : i \in I)$ *be an A-independent sequence, let* $\pi_i(x)$ *a partial type over* Aa_i *which does not fork over A and assume that whenever* $(b_i : i \in I)$ *is a sequence of realizations* $b_i \models \pi_i$ *then* $b_i \stackrel{\text{Ls}}{\equiv}_A b_j$ *for all* $i, j \in I$. *Then* $\bigcup_{i \in I} \pi_i(x)$ *does not fork over A.*
3. *Let* $(a_i : i \in I)$ *be an M-independent sequence, let* $\pi_i(x)$ *a partial type over* Ma_i *which does not fork over M and extends* $p(x) \in S(M)$. *Then* $\bigcup_{i \in I} \pi_i(x)$ *does not fork over M.*

PROOF. 1 follows from Corollary 10.9. For 2 we may assume $I = \omega$ and then using 1 it is easy to prove by induction that $\pi_0(x) \cup \dots \cup \pi_n(x)$ does not fork over A for all $n < \omega$. 3 follows from 2 since $b_i \equiv_M b_j$ implies $b_i \stackrel{\text{Ls}}{\equiv}_M b_j$. ⊣

PROPOSITION 10.11. *Let T be simple. If* $a \stackrel{\text{Ls}}{\equiv}_A b$ *and* $a \mathop{\smile\!\!\!\!\!\mid}_A b$, *then* a, b *start a Morley sequence* $(a_i : i < \omega)$ *over A.*

PROOF. Let $p = \text{tp}(ab/A)$. We prove first that for any cardinal κ there is an A-independent sequence $(a_i : i < \kappa)$ such that $\models p(a_i, a_j)$ for all $i < j < \kappa$.

Note that this implies $a_i \overset{\text{Ls}}{\equiv}_A a_j$. The sequence is constructed inductively starting with $a_0 = a$ and $a_1 = b$. We choose as a_i a realization of $\bigcup_{j<i} p(a_j, x)$ such that $a_i \mathop{\smile\!\!\!\!\!\!\;|}_A a_{<i}$. To do this we need to prove that $\bigcup_{j<i} p(a_j, x)$ does not fork over A. Note that $c \overset{\text{Ls}}{\equiv}_A d$ whenever $c \models p(a_j, x)$ and $d \models p(a_{j'}, x)$. Therefore it is clear that we can apply the generalized version of the Independence Theorem stated in point 2 of Corollary 10.10 to obtain the desired result. Now, once we have this A-independent sequence we still need to make it A-indiscernible. But this can be done easily by Proposition 1.6. ⊣

PROPOSITION 10.12. *If T is simple, then $a \overset{\text{Ls}}{\equiv}_A b$ if and only if there is some c such that a, c start an infinite A-indiscernible sequence and b, c start an infinite indiscernible sequence over A.*

PROOF. Assume $a \overset{\text{Ls}}{\equiv}_A b$ and find with Lemma 10.2 some c such that $c \overset{\text{Ls}}{\equiv}_A a$ and $c \mathop{\smile\!\!\!\!\!\!\;|}_A ab$. By Proposition 10.11 a, c start an infinite Morley sequence over A and b, c start an infinite Morley sequence over A. ⊣

COROLLARY 10.13. *If T is simple, then the relation $\overset{\text{Ls}}{\equiv}_A$ of equality of Lascar strong type over A is type-definable over A by $\mathrm{nc}_A^2(x, y)$.*

PROOF. Clear, by Proposition 10.12. ⊣

COROLLARY 10.14. *If T is simple, then for any set A, $a \overset{\text{Ls}}{\equiv}_A b$ if and only if $a \overset{\text{Ls}}{\equiv}_{A'} b$ for all finite $A' \subseteq A$.*

PROOF. By Corollary 10.13. ⊣

DEFINITION 10.15. T is *G-compact over A* if the relation $\overset{\text{Ls}}{\equiv}_A$ of equality of Lascar strong type is type-definable for every possible length of tuples.

REMARK 10.16. *The following are equivalent:*

1. *T is G-compact over A.*
2. $\overset{\text{Ls}}{\equiv}_A = \overset{\text{KP}}{\equiv}_A$.
3. *For some $n < \omega$, $\overset{\text{Ls}}{\equiv}_A$ is defined by the type $\mathrm{nc}_A^n(x, y)$.*
4. *For some $n < \omega$, all Lascar strong types over A have diameter $\leq n$.*

PROOF. Clearly both 2 and 3 imply 1. By Remark 9.11, 3 is equivalent to 4. For 1 ⇒ 2 note that if $\overset{\text{Ls}}{\equiv}_A$ is type-definable then it is the least type-definable over A bounded equivalence relation and hence it must be $\overset{\text{KP}}{\equiv}_A$. Finally for 1 ⇒ 3 use Corollary 9.22. ⊣

COROLLARY 10.17. *If T is simple, then T is G-compact over A for every A.*

PROOF. By Corollary 10.13, $\overset{\text{Ls}}{\equiv}_A$ is type-definable over A. ⊣

EXAMPLE 10.18. The first example of a non G-compact theory was discovered by Ziegler. It is presented in [10].

PROPOSITION 10.19. *T is G-compact over A if and only if the two following conditions hold:*

1. *For every $n < \omega$ the relation $\overset{\text{Ls}}{\equiv}_A$ of equality of Lascar strong type on n-tuples is type-definable.*
2. $\operatorname{Autf}(\mathfrak{C}/A)$ *is closed in the topology of* $\operatorname{Aut}(\mathfrak{C}/A)$.

PROOF. Assume T is G-compact over A. We need only to check 2. Let $f \in \operatorname{Aut}(\mathfrak{C}/A)$ be an accumulation point of $\operatorname{Autf}(\mathfrak{C}/A)$ and let m be a tuple enumerating a model $M \supseteq A$. For each finite subtuple a of m there is some $g \in \operatorname{Autf}(\mathfrak{C}/A)$ such that $g(a) = f(a)$ and hence by Corollary 9.15, $a \overset{\text{Ls}}{\equiv}_A f(a)$. Since $\overset{\text{Ls}}{\equiv}_A$ is type-definable over A, it follows $m \overset{\text{Ls}}{\equiv}_A f(m)$. Again by Corollary 9.15, there is some $g \in \operatorname{Autf}(\mathfrak{C}/A)$ such that $f(m) = g(m)$. Then $g^{-1} \circ f$ is the identity on M and therefore it belongs to $\operatorname{Autf}(\mathfrak{C}/A)$. It follows that $f = g \circ g^{-1} \circ f \in \operatorname{Autf}(\mathfrak{C}/A)$.

Assume now 1 and 2. We check that for each infinite ordinal α, $\overset{\text{Ls}}{\equiv}_A$ as a relation on tuples of length α is type-definable. Fix α-tuples of variables $x = (x_i : i < \alpha)$ and $y = (y_i : i < \alpha)$ and for each finite $I \subseteq \alpha$ choose a type $\Sigma_I(x_I, y_I)$ over A defining $\overset{\text{Ls}}{\equiv}_A$ on the corresponding finite tuples of variables $x_I = (x_i : i \in I)$ and $y_I = (y_i : i \in I)$. Let $\Sigma_\alpha(x, y)$ be the union of all these types. It is easy to see that $\models \Sigma_\alpha(a, b)$ if and only if the corresponding elementary mapping $a \mapsto b$ preserves $\overset{\text{Ls}}{\equiv}_A$ on finite subtuples. It is also clear that these elementary mappings can be extended by back-and-forth. Hence, if $\models \Sigma_\alpha(a, b)$ then $a \mapsto b$ can be extended to some $f \in \operatorname{Aut}(\mathfrak{C}/A)$ such that for every finite tuple c, $c \overset{\text{Ls}}{\equiv}_A f(c)$. This means that f is an accumulation point of $\operatorname{Autf}(\mathfrak{C}/A)$. By 2, $f \in \operatorname{Autf}(\mathfrak{C}/A)$ and therefore $a \overset{\text{Ls}}{\equiv}_A b$. ⊣

Chapter 11

CANONICAL BASES

DEFINITION 11.1. The *multiplicity of a type* $p(x) \in S(A)$ is the number $\mathrm{Mlt}(p)$ of its global nonforking extensions $\mathfrak{p}(x) \in S(\mathfrak{C})$. If there is a proper class of global nonforking extensions of p, we say that p has unbounded multiplicity and we write $\mathrm{Mlt}(p) = \infty$; otherwise we say that p has bounded multiplicity. A *stationary type* is a type of multiplicity 1. Thus over any $B \supseteq A$ a stationary type $p(x) \in S(A)$ has a unique nonforking extension $q(x) \in S(B)$. We use the notation $p|B$ for q.

LEMMA 11.2. *Let T be simple. If $p \in S(A)$ is stationary, then its global nonforking extension is definable over A.*

PROOF. Let $\mathfrak{p}$ be the global nonforking extension of p, and let $\varphi(x, y) \in L$. We will show that $\mathfrak{p} \restriction \varphi$ is A-definable. Let $\Delta_\varphi(y)$ and $\Delta_{\neg\varphi}(y)$ be types over A given by Corollary 5.23 for p and φ and for p and $\neg\varphi$ respectively. By compactness, the conjunction $\psi(y)$ of a finite subset of $\Delta_\varphi(y)$ is inconsistent with $\Delta_{\neg\varphi}(y)$. It is clear that $\psi(y)$ defines $\mathfrak{p} \restriction \varphi$. ⊣

COROLLARY 11.3. *Let T be simple. If types over models are stationary, then T is stable.*

PROOF. Lemma 11.2 implies that in this situation every global type is definable. ⊣

PROPOSITION 11.4. *Let T be simple.*

1. *If $p \in S(M)$ has bounded multiplicity, then p is stationary.*
2. *If $p \in S(A)$ has bounded multiplicity, then every extension of p over $\mathrm{acl}^{\mathrm{eq}}(A)$ is stationary.*

PROOF. 1. Assume $p \in S(M)$ has two nonforking extensions over $A \supseteq M$, say p_1 and p_2. We will show that no nonforking extension of p is stationary. This implies that p has an unbounded number of nonforking global extensions. Let q be a nonforking extension of p over $B \supseteq M$. To show that q is not stationary we may assume $B \mathop{\smile\!\!\!\!\!\!|}_M A$. By the Independence Theorem (Corollary 10.10) applied to p_1 and q we obtain a type $q_1 \in S(AB)$ extending $q \cup p_1$ which does not fork over M. Similarly, by applying it to p_2 and q we obtain a type $q_2 \in S(AB)$ extending $q \cup p_2$ which does not fork over M. Then

q_1, q_2 are two different nonforking extensions of q over AB, which shows that q is not stationary.

2. Let $p'(x) \in S(\mathrm{acl}^{\mathrm{eq}}(A))$ be a (nonforking) extension of p and let $M \supseteq A$. Any nonforking extension of p' over M has bounded multiplicity and by point 1 it is stationary. We show that p' has only one nonforking extension over M. This will ensure the stationarity of p'. Let $q_1 \in S(M)$ be a nonforking extension of p'. By Lemma 11.2 the global nonforking extension of q_1 is M-definable. Since p' has bounded multiplicity, this global nonforking extension has a bounded number of $\mathrm{acl}^{\mathrm{eq}}(A)$-conjugates and therefore it is definable over $\mathrm{acl}^{\mathrm{eq}}(A)$. Therefore q_1 is definable over $\mathrm{acl}^{\mathrm{eq}}(A)$. Now assume that $q_2 \in S(M)$ is another nonforking extension of p'. Again, q_2 is stationary and definable over $\mathrm{acl}^{\mathrm{eq}}(A)$.

Let $d_1 x\varphi(x, y) \in L(\mathrm{acl}^{\mathrm{eq}}(A))$ and $d_2 x\varphi(x, y) \in L(\mathrm{acl}^{\mathrm{eq}}(A))$ be definitions of q_1 and q_2 respectively. We will show that if $\varphi(x, a) \in q_1$ then $\varphi(x, a) \in q_2$. Let $b_i \models q_i$. Then $b_i \mathop{\smile\!\!\!\!\!\!\mid}_A M$ and $\models \varphi(b_1, a)$. Let $r(y) = \mathrm{stp}(a/A)$ and let $\Delta(x)$ be the partial type over $\mathrm{acl}^{\mathrm{eq}}(A)$ given by Corollary 5.23 for $r(y)$ and $\varphi^{-1}(y, x)$. Then $\models \Delta(b_1)$. Since it is a partial type over $\mathrm{acl}^{\mathrm{eq}}(A)$, also $\models \Delta(b_2)$ and therefore there is some a' such that $a' \mathop{\smile\!\!\!\!\!\!\mid}_A b_2$, $a' \models r(y)$ and $\models \varphi(b_2, a')$. We may find this a' with the additional property that $a' \mathop{\smile\!\!\!\!\!\!\mid}_{Ab_2} M$. In this case $a' \mathop{\smile\!\!\!\!\!\!\mid}_M b_2$ and hence by stationarity $\varphi(x, a')$ belongs to the global nonforking extension of q_2, that is, $\models d_2 x\varphi(x, a')$. Since this formula is over $\mathrm{acl}^{\mathrm{eq}}(A)$ and $a \stackrel{s}{\equiv}_A a'$ we conclude that $\models d_2 x\varphi(x, a)$, that is, $\varphi(x, a) \in q_2$. ⊣

REMARK 11.5. *Let T be stable.*

1. *Any strong type is stationary.*
2. *Any type over a model is stationary.*

PROOF. Clear by Proposition 9.28. ⊣

LEMMA 11.6. *Let T be simple.* $\mathrm{tp}(ab/A)$ *is stationary if and only if* $\mathrm{tp}(b/A)$ *and* $\mathrm{tp}(a/Ab)$ *are stationary.*

PROOF. Let $\mathrm{tp}(ab/A)$ be stationary. We check first that $\mathrm{tp}(b/A)$ is stationary. For this assume $B \supseteq A$, $b' \equiv_A b$, $b \mathop{\smile\!\!\!\!\!\!\mid}_A B$, and $b' \mathop{\smile\!\!\!\!\!\!\mid}_A B$. Choose a_0 such that $a_0 b \equiv_A ab$ and $a_0 \mathop{\smile\!\!\!\!\!\!\mid}_{Ab} B$ and then choose a_1 such that $a_1 b' \equiv_A ab$ and $a_1 \mathop{\smile\!\!\!\!\!\!\mid}_{Ab'} B$. Then $a_0 b \mathop{\smile\!\!\!\!\!\!\mid}_A B$ and $a_1 b' \mathop{\smile\!\!\!\!\!\!\mid}_A B$. Since $\mathrm{tp}(ab/A)$ is stationary, $a_0 b \equiv_B a_1 b'$ and therefore $b \equiv_B b'$. Now we check that $\mathrm{tp}(a/Ab)$ is stationary. Assume $a \equiv_{Ab} a'$, $B \supseteq A$, $a \mathop{\smile\!\!\!\!\!\!\mid}_{Ab} B$ and $a' \mathop{\smile\!\!\!\!\!\!\mid}_{Ab} B$. Let $b' \equiv_A b$ be such that $b' \mathop{\smile\!\!\!\!\!\!\mid}_A B$. We have just proved that $\mathrm{tp}(b/A)$ is stationary. Hence $b \equiv_B b'$. Choose now a_0, a_1 such that $a_0 a_1 b' \equiv_B aa'b$. Then $a_0 b' \mathop{\smile\!\!\!\!\!\!\mid}_A B$, $a_1 b' \mathop{\smile\!\!\!\!\!\!\mid}_A B$, and $a_0 b' \equiv_A a_1 b' \equiv_A ab$. By stationarity $a_0 b' \equiv_B a_1 b'$, and hence $ab \equiv_B a'b$ and $a \equiv_{Bb} a'$.

Now assume $\mathrm{tp}(b/A)$ and $\mathrm{tp}(a/Ab)$ are stationary. Let $ab \equiv_A a'b'$, $B \supseteq A$, $ab \mathop{\smile\!\!\!\!\!\!\mid}_A B$, and $a'b' \mathop{\smile\!\!\!\!\!\!\mid}_A B$. By stationarity of $\mathrm{tp}(b/A)$, $b \equiv_B b'$. Now choose

a_0 such that $a_0 b \equiv_B a'b'$. Then $a_0 \mathop{\smile\!\!\!\!\!\!\,|}_{Ab} B$ and $a_0 \equiv_{Ab} a$. By stationarity of $\mathrm{tp}(a/Ab)$, $a_0 \equiv_{Bb} a$. Hence $ab \equiv_B a_0 b \equiv_B a'b'$. ⊣

REMARK 11.7. *If T is stable, then any two global nonforking extensions of $p(x) \in S(A)$ are A-conjugate.*

PROOF. Let $\mathfrak{p}_1, \mathfrak{p}_2 \in S(\mathfrak{C})$ be two nonforking extensions of p and let $p_i = \mathfrak{p}_i \restriction \mathrm{acl}^{\mathrm{eq}}(A)$. There is some $f \in \mathrm{Aut}(\mathfrak{C}/A)$ such that $p_1^f = p_2$. By Remark 11.5, p_2 is stationary. Since $\mathfrak{p}_1^f$ and $\mathfrak{p}_2$ are nonforking extensions of p_2, they coincide. ⊣

PROPOSITION 11.8. *Let T be stable and let $p(x) \in S(A)$ be finitary.*

1. $\mathrm{Mlt}(p) \leq 2^{|T|}$.
2. *If* $\mathrm{Mlt}(p) \geq \omega$, *then* $\mathrm{Mlt}(p) \geq 2^{\omega}$.

PROOF. 1. Choose some $B \subseteq A$ of cardinality $\leq |T|$ such that p does not fork over B. Since every nonforking extension of p is a nonforking extension of $p \restriction B$, it is enough to check that $\mathrm{Mlt}(p \restriction B) \leq 2^{|T|}$. Let $M \supseteq B$ be a model of cardinality $\leq |T|$. By Remark 11.5 every type over M extending $p \restriction B$ is stationary. Then $\mathrm{Mlt}(p \restriction B)$ is bounded by the number of extensions of $p \restriction B$ over M and this number is $\leq |S(M)| \leq 2^{|T|}$.

2. By Corollary 8.11, the set of nonforking extensions over $\mathfrak{C}$ of $p(x) \in S(A)$ is a closed set in $S(\mathfrak{C})$. By Remark 11.7 any two points in this set are connected by a homeomorphism induced by an automorphism of $\mathfrak{C}$ over A. Hence if this set has an isolated point, any other point is isolated and therefore it is finite. If it does not have isolated points, it is a nonempty perfect set and therefore it contains at least $\geq 2^{\omega}$ points. ⊣

DEFINITION 11.9. Two stationary types $p(x) \in S(A)$, $q(x) \in S(B)$ are called *parallel* if they have a common nonforking extension. We then write $p \parallel q$. Note that $q = (p|AB) \restriction B$.

DEFINITION 11.10. Let $\mathfrak{p} \in S(\mathfrak{C})$ be definable. A subset B of $\mathfrak{C}^{\mathrm{eq}}$ is a *canonical base* of $\mathfrak{p}$ if for every $f \in \mathrm{Aut}(\mathfrak{C})$, $\mathfrak{p}^f = \mathfrak{p}$ if and only if f fixes B pointwise. Clearly, $\mathfrak{p}$ is definable over A if and only if $B \subseteq \mathrm{dcl}^{\mathrm{eq}}(A)$. Hence, in a sense, a canonical base of $\mathfrak{p}$ is a smallest set over which $\mathfrak{p}$ is definable.

REMARK 11.11. 1. *If $\mathfrak{p}$ is definable and B is a canonical base of $\mathfrak{p}$, then $\mathfrak{p}$ is definable over B.*

2. *If B, B' are canonical bases of the definable type $\mathfrak{p}$, then* $\mathrm{dcl}^{\mathrm{eq}}(B) = \mathrm{dcl}^{\mathrm{eq}}(B')$.
3. *For every definable global type $\mathfrak{p}$, if c_φ is the canonical parameter of some definition of $\mathfrak{p} \restriction \varphi$, then $(c_\varphi : \varphi \in L)$ is a canonical base of $\mathfrak{p}$.*

PROOF. Easy exercise. ⊣

DEFINITION 11.12. Let $p(x) \in S(A)$ be a stationary type in a simple theory. We call B a *canonical base* of p if B is a canonical base of the (definable) global nonforking extension of p. We use the notation $\mathrm{Cb}(p)$ for $\mathrm{dcl}^{\mathrm{eq}}(B)$ where B is a canonical base of p. Finally, in a stable theory we define

$$\mathrm{Cb}(a/A) = \mathrm{Cb}(\mathrm{stp}(a/A)).$$

REMARK 11.13. *Let T be simple. A set B is a canonical base of the stationary type $p \in S(A)$ if and only if for each $f \in \mathrm{Aut}(\mathfrak{C})$: $p \parallel p^f$ if and only if f fixes B pointwise.*

PROOF. Clear, since an automorphism f fixes the global nonforking extension of p if and only if $p \parallel p^f$. ⊣

PROPOSITION 11.14. *Let T be simple. If $p(x) \in S(A)$ is stationary, then $\mathrm{Cb}(p) \subseteq \mathrm{dcl}^{\mathrm{eq}}(A)$. If T is stable, then $\mathrm{Cb}(a/A) \subseteq \mathrm{acl}^{\mathrm{eq}}(A)$.*

PROOF. If $f \in \mathrm{Aut}(\mathfrak{C}/A)$ and $p(x) \in S(A)$ is stationary, then $p^f = p \parallel p$ and therefore f fixes $\mathrm{Cb}(p)$ pointwise. Hence $\mathrm{Cb}(p) \subseteq \mathrm{dcl}^{\mathrm{eq}}(A)$. Now note that in a stable theory $\mathrm{Cb}(a/A) = \mathrm{Cb}(\mathrm{stp}(a/A))$ and $\mathrm{stp}(a/A) = \mathrm{tp}(a/\mathrm{acl}^{\mathrm{eq}}(A))$ is stationary. ⊣

PROPOSITION 11.15. *Let T be stable. Let B be a canonical base of $\mathfrak{p}(x) \in S(\mathfrak{C})$. Then $\mathfrak{p}$ does not fork over A if and only if $B \subseteq \mathrm{acl}^{\mathrm{eq}}(A)$. Moreover the following are equivalent:*

1. *$\mathfrak{p}$ is definable over A.*
2. *$B \subseteq \mathrm{dcl}^{\mathrm{eq}}(A)$.*
3. *$\mathfrak{p}$ does not fork over A and $\mathfrak{p} \upharpoonright A$ is stationary.*

PROOF. If $\mathfrak{p}$ does not fork over A then $p(x) = \mathfrak{p} \upharpoonright \mathrm{acl}^{\mathrm{eq}} A$ is stationary and has B as a canonical base. Hence by Proposition 11.14 $B \subseteq \mathrm{acl}^{\mathrm{eq}}(A)$. On the other hand if $B \subseteq \mathrm{acl}^{\mathrm{eq}} A$ then $\mathfrak{p}$ is definable over $\mathrm{acl}^{\mathrm{eq}}(A)$ and hence it does not fork over A.

Equivalence between 1 and 2 is immediate. Now we prove the equivalence with 3. If $\mathfrak{p}$ is definable over A, then by Corollary 8.7 $\mathfrak{p}$ does not fork over A. Moreover $\mathfrak{p}$ is the only element of its orbit in $\mathrm{Aut}(\mathfrak{C}/A)$ and then, by Remark 11.7, $\mathfrak{p} \upharpoonright A$ is stationary. For the other direction, if $\mathfrak{p}$ does not fork over A and $\mathfrak{p} \upharpoonright A$ is stationary, then clearly $\mathfrak{p}$ is the only element of its orbit in $\mathrm{Aut}(\mathfrak{C}/A)$ and therefore it is A-definable. ⊣

PROPOSITION 11.16. *Let T be stable. If $A \subseteq B$, the following are equivalent.*

1. $a \mathop{\smile\!\!\!\!\!\!\mid}_A B$.
2. $\mathrm{Cb}(a/B) \subseteq \mathrm{acl}^{\mathrm{eq}}(A)$.
3. $\mathrm{Cb}(a/A) = \mathrm{Cb}(a/B)$.

PROOF. $1 \Leftrightarrow 2$ follows from Proposition 11.15. As for 3, note that if $a \mathop{\smile\!\!\!\!\!\!\mid}_A B$ then $\mathrm{stp}(a/A)$ and $\mathrm{stp}(a/B)$ have the same global nonforking extension and therefore $\mathrm{Cb}(a/A) = \mathrm{Cb}(a/B)$. On the other hand, if their canonical bases coincide, then $\mathrm{Cb}(a/B) = \mathrm{Cb}(a/A) \subseteq \mathrm{acl}^{\mathrm{eq}}(A)$. ⊣

LEMMA 11.17. *Let T be stable. If a is a tuple definable over the tuple b, then* $\mathrm{Cb}(a/A) \subseteq \mathrm{Cb}(b/A)$. *Hence, two interdefinable tuples have the same canonical base over any set.*

PROOF. Let $C = \mathrm{Cb}(b/A)$. Since $b \mathop{\smile\!\!\!\!\!\mid}_C A$ and $\mathrm{tp}(b/C)$ is stationary then $a \mathop{\smile\!\!\!\!\!\mid}_C A$ and by Lemma 11.6 $\mathrm{tp}(a/C)$ is stationary. Hence $\mathrm{Cb}(a/A) \subseteq C$. ⊣

LEMMA 11.18. *Let T be simple. Let $p(x), q(y) \in S(A)$ and assume one of them is stationary. Let a, a' be realizations of p and let b, b' be realizations of q. If $a \mathop{\smile\!\!\!\!\!\mid}_A b$ and $a' \mathop{\smile\!\!\!\!\!\mid}_A b'$ then $ab \equiv_A a'b'$. If p and q are stationary, then* $\mathrm{tp}(ab/A)$ *is also stationary.*

PROOF. Without loss of generality, q is stationary. Choose c such that $ab \equiv_A a'c$. Then $c \equiv_A b'$, $c \mathop{\smile\!\!\!\!\!\mid}_A a'$ and $b' \mathop{\smile\!\!\!\!\!\mid}_A a'$. Since q is stationary, $c \equiv_{Aa'} b'$. Then $a'b' \equiv_A a'c \equiv_A ab$. The last assertion follows from Lemma 11.6. ⊣

LEMMA 11.19. *Let T be simple. Let $(I, <)$ be a linearly ordered set and for each $i \in I$, let $p_i(x_i) \in S(A)$ be stationary. Let $(a_i : i \in I)$ be an A-independent sequence where $a_i \models p_i$ for all $i \in I$. If $(b_i : i \in I)$ is an A-independent sequence such that $b_i \models p_i$ for all $i \in I$, then $(a_i : i \in I) \equiv_A (b_i : i \in I)$. Moreover* $\mathrm{tp}((a_i : i \in I)/A)$ *is stationary.*

PROOF. We can assume I is finite and then it can be proved easily by induction on $|I|$ using Lemma 11.18. ⊣

DEFINITION 11.20. Assume T is simple. Let $p_i(x_i) \in S(A)$ for each $i \in I$ and assume each of the types p_i is stationary. The *product* of the types $(p_i : i \in I)$ is the stationary type $\mathrm{tp}((a_i : i \in I)/A)$ where $(a_i : i \in I)$ is A-independent and $a_i \models p_i$. By Lemma 11.19 it is well defined. We denote it by $\bigotimes_{i \in I} p_i$. In the finite case we use the notation $p_1 \otimes \cdots \otimes p_n$. If all the types p_i are equal to $p(x) \in S(A)$, the notations are p^I and p^n.

REMARK 11.21. *Let T be simple. If $(a_i : i < \alpha)$ is an A-independent sequence of realizations of the stationary type $p(x) \in S(A)$, then it is a Morley sequence in p and* $\mathrm{tp}((a_i : i < \alpha)/A) = p^\alpha$. *Hence, if $(b_i : i < \alpha)$ is another A-independent sequence of realizations of p, then $(a_i : i < \alpha) \equiv_A (b_i : i < \alpha)$.*

PROOF. A-indiscernibility of $(a_i : i < \alpha)$ can be justified noticing that for every $n < \omega$, for every $i_0 < \cdots < i_n < \alpha$,

$$\mathrm{tp}(a_0, \ldots, a_n/A) = \mathrm{tp}(a_{i_0}, \ldots, a_{i_n}/A) = p^{n+1}. \qquad ⊣$$

LEMMA 11.22. *Let $\varphi(x, y) \in L$ stable, let $\mathfrak{p} \in S(\mathfrak{C})$ and assume $\mathfrak{p} \restriction \varphi$ is M-definable. If c_φ is the canonical parameter of some definition of $\mathfrak{p} \restriction \varphi$ over M, then $c_\varphi \in \mathrm{dcl}^{\mathrm{eq}}(a_i : i < \omega)$ for some Morley sequence $(a_i : i < \omega)$ in $\mathfrak{p} \restriction M$.*

PROOF. By Proposition 8.2 $\mathfrak{p} \restriction \varphi$ is definable over some Morley sequence $(a_i : i < \omega)$ in $\mathfrak{p} \restriction M$. ⊣

PROPOSITION 11.23. *If T is stable, then for each Morley sequence $(a_i : i < \omega)$ in* $\mathrm{stp}(a/A)$, $\mathrm{Cb}(a/A) \subseteq \mathrm{dcl}^{\mathrm{eq}}(a_i : i < \omega)$.

PROOF. Let $\mathfrak{p}$ be the global nonforking extension of $p(x) = \mathrm{stp}(a/A)$ and fix some $\varphi(x, y) \in L$ and some model $M \supseteq A$. Let c_φ be the canonical parameter of a definition of $\mathfrak{p} \restriction \varphi$. By Lemma 11.22 $c_\varphi \in \mathrm{dcl}^{\mathrm{eq}}(b_i : i < \omega)$ for some Morley sequence $(b_i : i < \omega)$ in $\mathfrak{p} \restriction M$. Note that $(b_i : i < \omega)$ is also a Morley sequence in $\mathrm{stp}(a/A)$. By Remark 11.21 $(a_i : i < \omega) \overset{\mathrm{s}}{\equiv}_A (b_i : i < \omega)$ and therefore there is some $f \in \mathrm{Aut}(\mathfrak{C}/\mathrm{acl}^{\mathrm{eq}}(A))$ sending each b_i to a_i. Since $\mathfrak{p}$ is definable over $\mathrm{acl}^{\mathrm{eq}}(A)$, $c_\varphi \in \mathrm{acl}^{\mathrm{eq}}(A)$. It follows that $c_\varphi \in \mathrm{dcl}^{\mathrm{eq}}(a_i : i < \omega)$. Since $\mathrm{Cb}(a/A)$ is definable over $(c_\varphi : \varphi \in L)$, we conclude that $\mathrm{Cb}(a/A) \subseteq \mathrm{dcl}^{\mathrm{eq}}(a_i : i < \omega)$. ⊣

Chapter 12

ABSTRACT INDEPENDENCE RELATIONS

NOTATION 12.1. *In this chapter* $\mathop{\smile\!\!\!\!\!\mid}$ *will be an arbitrary ternary relation between sets. We will use* $\mathop{\smile\!\!\!\!\!\mid}^f$ *for the nonforking independence relation as defined in* 5.1.

DEFINITION 12.2. An *independence relation*is a ternary relation $\mathop{\smile\!\!\!\!\!\mid}$ between sets satisfying the following axioms:

1. Invariance. If $A \mathop{\smile\!\!\!\!\!\mid}_C B$ and $f \in \mathrm{Aut}(\mathfrak{C})$, then $f(A) \mathop{\smile\!\!\!\!\!\mid}_{f(C)} f(B)$.
2. Monotonicity. If $A \mathop{\smile\!\!\!\!\!\mid}_C B$, $A' \subseteq A$, and $B' \subseteq B$, then $A' \mathop{\smile\!\!\!\!\!\mid}_C B'$.
3. Right base monotonicity. If $A \mathop{\smile\!\!\!\!\!\mid}_C B$ and $C \subseteq D \subseteq B$, then $A \mathop{\smile\!\!\!\!\!\mid}_D B$.
4. Left transitivity. If $B \subseteq C \subseteq D$, $C \mathop{\smile\!\!\!\!\!\mid}_B A$, and $D \mathop{\smile\!\!\!\!\!\mid}_C A$, then $D \mathop{\smile\!\!\!\!\!\mid}_B A$.
5. Left normality. If $A \mathop{\smile\!\!\!\!\!\mid}_C B$, then $AC \mathop{\smile\!\!\!\!\!\mid}_C B$.
6. Extension. If $A \mathop{\smile\!\!\!\!\!\mid}_C B$ and $B' \supseteq B$, then $A' \mathop{\smile\!\!\!\!\!\mid}_C B'$ for some $A' \equiv_{BC} A$.
7. Left finite character. If $A_0 \mathop{\smile\!\!\!\!\!\mid}_C B$ for all finite $A_0 \subseteq A$, then $A \mathop{\smile\!\!\!\!\!\mid}_C B$.
8. Local character. For every A there is a cardinal number $\kappa(A)$ such that for any B there is some $C \subseteq B$ such that $|C| < \kappa(A)$ and $A \mathop{\smile\!\!\!\!\!\mid}_C B$.

We say that the independence relation $\mathop{\smile\!\!\!\!\!\mid}$ is *strict* if it additionally satisfies

9. Anti-reflexivity. If $A \mathop{\smile\!\!\!\!\!\mid}_C A$, then $A \subseteq \mathrm{acl}(C)$.

For a tuple a, $a \mathop{\smile\!\!\!\!\!\mid}_C B$ means that $A \mathop{\smile\!\!\!\!\!\mid}_C B$ where A is the set enumerated by a. Similarly for other notations like $a \mathop{\smile\!\!\!\!\!\mid}_C b$, etc.

REMARK 12.3. *Note that the property of* right normality

$$\text{if } A \mathop{\smile\!\!\!\!\!\mid}_C B \text{ then } A \mathop{\smile\!\!\!\!\!\mid}_C BC$$

follows from extension and invariance. In this context, right base monotonicity may be reformulated as:

$$\text{if } A \mathop{\smile\!\!\!\!\!\mid}_C B \text{ and } D \subseteq B, \text{ then } A \mathop{\smile\!\!\!\!\!\mid}_{CD} B.$$

Similarly, under left normality, left transitivity can be reformulated as:

$$\text{if } C \mathop{\smile\!\!\!\!\!\mid}_B A, \text{ and } D \mathop{\smile\!\!\!\!\!\mid}_{BC} A, \text{ then } CD \mathop{\smile\!\!\!\!\!\mid}_B A.$$

In this form the property is one direction of the Pairs Lemma:

$$CD \underset{B}{\perp} A \text{ if and only if } C \underset{B}{\perp} A \text{ and } D \underset{BC}{\perp} A.$$

Note also that right base monotonicity and weak local character give what is known as the existence property:

$$A \underset{B}{\perp} B.$$

PROPOSITION 12.4. *Assume $\perp$ satisfies the first five axioms of independence relations and also the extension property. If $a \perp_C B$, then there is a BC-indiscernible sequence $(a_i : i < \omega)$ such that $a_i \equiv_{BC} a$ and $a_{<i} \perp_C a_i$ for all $i < \omega$.*

PROOF. We will use the monotonicity property several times without explicitly mentioning it. Since $a \perp_C B$, by the extension property for any λ we can construct a sequence $(a_i : i < \lambda)$ such that $a_0 = a$, $a_i \equiv_{BC} a_0$, and $a_i \perp_C Ba_{<i}$. If we choose λ large enough and we apply Proposition 1.6, we obtain a BC-indiscernible sequence $(a'_i : i < \omega)$ such that for each $n < \omega$ there are $i_0 < \cdots < i_n < \lambda$ such that $a'_0, \dots, a'_n \equiv_{BC} a_{i_0}, \dots, a_{i_n}$. By invariance, $a'_i \perp_C a'_{<i}$ for all $i < \omega$. We now claim that for all $n > 0$,

$$(a'_i : 0 < i < n) \underset{C}{\perp} a'_0.$$

We prove it by induction on n. It is clear for $n = 1$. By the induction hypothesis and left normality, $C(a'_i : 0 < i < n) \perp_C a'_0$. By construction of the sequence and right base monotonicity, $a'_n \perp_{C(a'_i:0<i<n)} a'_0$. By left normality again, $C(a'_i : 0 < i \leq n) \perp_{C(a'_i:0<i<n)} a'_0$. Finally by left transitivity $C(a'_i : 0 < i \leq n) \perp_C a'_0$ and then $(a'_i : 0 < i \leq n) \perp_C a'_0$. This finishes the induction.

By compactness, there is a sequence $(a''_i : i < \omega)$ such that for each $n < \omega$, $a''_0, \dots, a''_n \equiv_{BC} a'_n, \dots, a'_0$. It is clear that it satisfies the required conditions. ⊣

PROPOSITION 12.5. *Assume $\perp$ satisfies the first five axioms of independence relations and also local character and finite character properties. Assume there is a BC-indiscernible sequence $(a_i : i < \omega)$ such that $a_i \equiv_{BC} a$ and $a_{<i} \perp_C a_i$ for all $i < \omega$. Then $B \perp_C a$.*

PROOF. Let $\kappa(B)$ be the cardinal given for B by the local character property and choose a regular cardinal $\kappa > \kappa(B)$. We can extend our sequence to a BC-indiscernible sequence $(a_i : i < \kappa)$. By finite character and invariance, $a_{<i} \perp_C a_i$ for all $i < \kappa$. By local character there is some $D \subseteq C \cup \{a_i : i < \kappa\}$ such that $|D| < \kappa$ and $B \perp_D C(a_i : i < \kappa)$. By regularity of κ, $D \subseteq Ca_{<j}$ for some $j < \kappa$. By right base monotonicity, $B \perp_{Ca_{<j}} C(a_i : i < \kappa)$ and by monotonicity, $B \mathop{\smile\!\!\!\!\!\!|}_{Ca_{<j}} a_j$. By left normality $BCa_{<j} \mathop{\smile\!\!\!\!\!\!|}_{Ca_{<j}} a_j$ and also $Ca_{<j} \mathop{\smile\!\!\!\!\!\!|}_C a_j$. By left transitivity, $BCa_{<j} \mathop{\smile\!\!\!\!\!\!|}_C a_j$. By monotonicity $B \mathop{\smile\!\!\!\!\!\!|}_C a_j$. Since $a \equiv_{BC} a_j$, by invariance $B \mathop{\smile\!\!\!\!\!\!|}_C a$. ⊣

COROLLARY 12.6. *Any independence relation is symmetric, that is: if* $A \mathop{\smile\!\!\!\!\!\!|}_C B$*, then* $B \mathop{\smile\!\!\!\!\!\!|}_C A$.

PROOF. It is an immediate consequence of propositions 12.4 and 12.5. ⊣

DEFINITION 12.7. For any invariant ternary relation $\mathop{\smile\!\!\!\!\!\!|}$ we define $\mathop{\smile\!\!\!\!\!\!|}^*$ as follows: $A \mathop{\smile\!\!\!\!\!\!|}^*_C B$ if and only if for all $B' \supseteq B$ there is some $A' \equiv_{BC} A$ such that $A' \mathop{\smile\!\!\!\!\!\!|}_C B'$. Sometimes it is more convenient to state it in terms of automorphisms: for all $B' \supseteq B$ there is some $f \in \mathrm{Aut}(\mathfrak{C}/BC)$ such that $f(A) \mathop{\smile\!\!\!\!\!\!|}_C B'$.

REMARK 12.8. *For any invariant* $\mathop{\smile\!\!\!\!\!\!|}$, $\mathop{\smile\!\!\!\!\!\!|}^*$ *is also invariant and* $A \mathop{\smile\!\!\!\!\!\!|}^*_C B$ *implies* $A \mathop{\smile\!\!\!\!\!\!|}_C B$.

PROPOSITION 12.9. *For any monotone invariant* $\mathop{\smile\!\!\!\!\!\!|}$, $\mathop{\smile\!\!\!\!\!\!|}^*$ *has the extension property.*

PROOF. Let $A \mathop{\smile\!\!\!\!\!\!|}^*_C B$ and $B \subseteq B'$. Let a enumerate A and let x be a corresponding tuple of variables. We claim that there is a type $p(x) \in S(CB')$ extending $\mathrm{tp}(a/CB)$ and such that for each cardinal κ there is a κ-saturated model $M \supseteq CB'$ and some $a' \models p$ such that $a' \mathop{\smile\!\!\!\!\!\!|}_C M$. Assume not, and fix for each $p(x) \in S(CB')$ extending $\mathrm{tp}(a/CB)$ a corresponding cardinal κ_p for which there is no κ_p-saturated model $M \supseteq CB'$ with a realization $a' \models p$ such that $a' \mathop{\smile\!\!\!\!\!\!|}_C M$. Let κ be the supremum of all these cardinals κ_p and choose a κ-saturated model $M \supseteq CB'$. Since $a \mathop{\smile\!\!\!\!\!\!|}^*_C B$, there is some $a' \equiv_{CB} a$ such that $a' \mathop{\smile\!\!\!\!\!\!|}_C M$. Then $\mathrm{tp}(a'/CB')$ provides a contradiction.

Now using the claim we fix some $p(x) \in S(CB')$ as indicated. Let $a' \models p$. We will show that $a' \mathop{\smile\!\!\!\!\!\!|}^*_C B'$. This will establish the extension property for $\mathop{\smile\!\!\!\!\!\!|}^*$. Let $B'' \supseteq B'$. We need to show that for some $a'' \equiv_{CB'} a'$ (i.e., some $a'' \models p$), $a'' \mathop{\smile\!\!\!\!\!\!|}_C B''$. Let $\kappa = |C \cup B'|^+ + |B''| + \omega$ and by the claim choose a κ-saturated $M \supseteq CB'$ and some $a'' \models p$ such that $a'' \mathop{\smile\!\!\!\!\!\!|}_C M$. By κ-saturation there is an automorphism $f \in \mathrm{Aut}(\mathfrak{C}/CB')$ such that $f(B'') \subseteq M$. By monotonicity $a'' \mathop{\smile\!\!\!\!\!\!|}_C f(B'')$. By invariance $f^{-1}(a'') \mathop{\smile\!\!\!\!\!\!|}_C B''$. Since $f^{-1}(a'') \models p$ we have finished. ⊣

REMARK 12.10. *Each one of the properties of monotonicity, right base monotonicity, left transitivity, left normality, and anti-reflexivity is preserved when passing from* $\mathop{\smile\!\!\!\!\!\!|}$ *to* $\mathop{\smile\!\!\!\!\!\!|}^*$.

PROPOSITION 12.11. *Assume* $\mathop{\smile\!\!\!\!\!\!|}$ *satisfies the first five axioms of independence and also left finite character. If* $\mathop{\smile\!\!\!\!\!\!|}^*$ *satisfies local character, then* $\mathop{\smile\!\!\!\!\!\!|}^*$ *is an independence relation.*

PROOF. By Remark 12.10 and Proposition 12.9 we only need to show that $\mathop{\smile\!\!\!\!\!\!\,|}^{*}$ has left finite character. But first we check that $\mathop{\smile\!\!\!\!\!\!\,|}^{*}$ is symmetric. Note that $\mathop{\smile\!\!\!\!\!\!\,|}^{*}$ satisfies the hypotheses of Proposition 12.4 and $\mathop{\smile\!\!\!\!\!\!\,|}$ satisfies the hypotheses of Proposition 12.5. Hence $A \mathop{\smile\!\!\!\!\!\!\,|}^{*}_{C} B$ implies $B \mathop{\smile\!\!\!\!\!\!\,|}_{C} A$.

Now assume $A \mathop{\smile\!\!\!\!\!\!\,|}^{*}_{C} B$ and let us prove that $B \mathop{\smile\!\!\!\!\!\!\,|}^{*}_{C} A$. Let $A' \supseteq A$. Since $A' \mathop{\smile\!\!\!\!\!\!\,|}^{*}_{AC} AC$, by extension there is some $f \in \mathrm{Aut}(\mathfrak{C}/AC)$ such that $f(A') \mathop{\smile\!\!\!\!\!\!\,|}^{*}_{AC} ACB$. By monotonicity $f(A') \mathop{\smile\!\!\!\!\!\!\,|}^{*}_{AC} B$. Since $A \mathop{\smile\!\!\!\!\!\!\,|}^{*}_{C} B$, by left transitivity and monotonicity of $\mathop{\smile\!\!\!\!\!\!\,|}^{*}$, $f(A') \mathop{\smile\!\!\!\!\!\!\,|}^{*}_{C} B$. Hence $B \mathop{\smile\!\!\!\!\!\!\,|}_{C} f(A')$ and $f^{-1}(B) \mathop{\smile\!\!\!\!\!\!\,|}_{C} A'$, which shows that $B \mathop{\smile\!\!\!\!\!\!\,|}^{*}_{C} A$.

Now we check left finite character of $\mathop{\smile\!\!\!\!\!\!\,|}^{*}$. Assume $a \mathop{\smile\!\!\!\!\!\!\,|}^{*}_{C} B$ for all finite tuples $a \in A$. To prove that $A \mathop{\smile\!\!\!\!\!\!\,|}^{*}_{C} B$, consider some $B' \supseteq B$. By existence and extension, there is some $f \in \mathrm{Aut}(\mathfrak{C}/BC)$ such that $f(A) \mathop{\smile\!\!\!\!\!\!\,|}^{*}_{CB} B'$. Hence $A \mathop{\smile\!\!\!\!\!\!\,|}^{*}_{CB} f^{-1}(B')$. By symmetry $f^{-1}(B') \mathop{\smile\!\!\!\!\!\!\,|}^{*}_{BC} A$. For each finite tuple $a \in A$, we have $a \mathop{\smile\!\!\!\!\!\!\,|}^{*}_{C} B$ and $a \mathop{\smile\!\!\!\!\!\!\,|}^{*}_{BC} f^{-1}(B')$. By symmetry and left transitivity we then obtain $a \mathop{\smile\!\!\!\!\!\!\,|}^{*}_{C} f^{-1}(B')$ for all finite tuples $a \in A$. Hence $a \mathop{\smile\!\!\!\!\!\!\,|}_{C} f^{-1}(B')$ for all finite tuples $a \in A$. By left finite character of $\mathop{\smile\!\!\!\!\!\!\,|}$, $A \mathop{\smile\!\!\!\!\!\!\,|}_{C} f^{-1}(B')$. By invariance $f(A) \mathop{\smile\!\!\!\!\!\!\,|}_{C} B'$. ⊣

PROPOSITION 12.12. *Let* $\mathop{\smile\!\!\!\!\!\!\,|}$ *be invariant and monotone. Then* $\mathop{\smile\!\!\!\!\!\!\,|} = \mathop{\smile\!\!\!\!\!\!\,|}^{*}$ *if and only if* $\mathop{\smile\!\!\!\!\!\!\,|}$ *has the extension property.*

PROOF. One direction follows from Proposition 12.9. The other direction is clear by definition of $\mathop{\smile\!\!\!\!\!\!\,|}^{*}$ since $\mathop{\smile\!\!\!\!\!\!\,|}^{*}$ is stronger than $\mathop{\smile\!\!\!\!\!\!\,|}$. ⊣

DEFINITION 12.13. It has already been mentioned that $\mathop{\smile\!\!\!\!\!\!\,|}^{f}$ is nonforking independence. We define $\mathop{\smile\!\!\!\!\!\!\,|}^{d}$ as nondividing independence. To be precise:

1. $A \mathop{\smile\!\!\!\!\!\!\,|}^{d}_{C} B$ if and only if for every tuple $a \in A$, $\mathrm{tp}(a/BC)$ does not divide over C.
2. $A \mathop{\smile\!\!\!\!\!\!\,|}^{f}_{C} B$ if and only if for every tuple $a \in A$, $\mathrm{tp}(a/BC)$ does not fork over C.

PROPOSITION 12.14. $(\mathop{\smile\!\!\!\!\!\!\,|}^{d})^{*} = \mathop{\smile\!\!\!\!\!\!\,|}^{f}$.

PROOF. By Remark 4.4 we know that $\mathop{\smile\!\!\!\!\!\!\,|}^{f}$ has the extension property. Since $\mathop{\smile\!\!\!\!\!\!\,|}^{f}$ implies $\mathop{\smile\!\!\!\!\!\!\,|}^{d}$, it follows that $\mathop{\smile\!\!\!\!\!\!\,|}^{f}$ implies $(\mathop{\smile\!\!\!\!\!\!\,|}^{d})^{*}$. For the other direction, assume $A (\mathop{\smile\!\!\!\!\!\!\,|}^{d})^{*}_{C} B$ but $A \not\mathop{\smile\!\!\!\!\!\!\,|}^{f}_{C} B$. For some tuple $a \in A$, for some formula $\varphi(x, y) \in L$, for some $b \in BC$, $\models \varphi(a, b)$ and $\varphi(x, b)$ forks over C. Then for some $\psi_1(x, y_1), \dots, \psi_n(x, y_n) \in L$, for some $b_1, \dots, b_n$,

$$\models \varphi(x, b) \rightarrow \psi_1(x, b_1) \vee \dots \vee \psi_n(x, b_n)$$

and each $\psi(x, b_i)$ divides over C. Let $B' = Bb_1, \dots, b_n$. By assumption there is some $a' \equiv_{BC} a$ such that $a' \mathop{\smile\hskip-0.9em ^| \ }^d_C B'$. Since $\models \varphi(a', b)$, $\models \psi_i(a', b_i)$ for some i. This implies that $\mathrm{tp}(a'/B')$ divides over C, a contradiction. ⊣

REMARK 12.15. *The relation $\mathop{\smile\hskip-0.9em ^| \ }^d$ has the properties of invariance, monotonicity, right base monotonicity, left transitivity, left normality, and anti-reflexivity. Therefore $\mathop{\smile\hskip-0.9em ^| \ }^f$ satisfies all these properties and extension. Moreover $\mathop{\smile\hskip-0.9em ^| \ }^d$ and $\mathop{\smile\hskip-0.9em ^| \ }^f$ satisfy left and right finite character.*

PROOF. For left transitivity see Proposition 4.8 and for anti-reflexivity see point 5 in Remark 4.2. The other properties are straightforward. ⊣

PROPOSITION 12.16. *The following are equivalent.*

1. *T is simple.*
2. *$\mathop{\smile\hskip-0.9em ^| \ }^f$ satisfies local character.*
3. *$\mathop{\smile\hskip-0.9em ^| \ }^d$ satisfies local character.*
4. *$\mathop{\smile\hskip-0.9em ^| \ }^f$ is an independence relation.*
5. *$\mathop{\smile\hskip-0.9em ^| \ }^d$ is an independence relation.*

PROOF. We know that simplicity of T implies all the other conditions. It is clear that 4 implies 2 and that 5 implies 3. It is also clear that 2 implies 3. We now check that simplicity follows from 3. Assume T is not simple. Then for some $p(x) \in S(\emptyset)$ for some $\varphi(x, y) \in L$, for some $k < \omega$, $\mathrm{D}(p(x), \varphi, k) = \infty$. The cardinal $\kappa(a)$ given by local character of $\mathop{\smile\hskip-0.9em ^| \ }^d$ is clearly the same for any realization a of p. Let κ be regular and larger than this cardinal. By Proposition 3.11 there is a sequence $(a_i : i < \kappa)$ such that $p(x) \cup \{\varphi(x, a_i) : i < \kappa\}$ is consistent and for each $i < \kappa$, $\varphi(x, a_i)$ k-divides over $a_{<i}$. Let $a \models p(x) \cup \{\varphi(x, a_i) : i < \kappa\}$. By choice of κ, there is some $C \subseteq \{a_i : i < \kappa\}$ such that $|C| < \kappa$ and $\mathrm{tp}(a/a_{<i})$ does not divide over C. By regularity of κ, for some $i < \kappa$, $C \subseteq \{a_j : j < i\}$. Then $\mathrm{tp}(a/a_{\leq i})$ does not divide over $a_{<i}$. But this contradicts the fact that $\models \varphi(a, a_i)$ and that $\varphi(x, a_i)$ divides over $a_{<i}$. ⊣

REMARK 12.17. *Assume $\mathop{\smile\hskip-0.9em ^| \ }$ is invariant and has local character. Let α be an ordinal number. There is a cardinal number κ such that for every tuple a of length α, for every set B there is some $C \subseteq B$ such that $|C| < \kappa$ and $a \mathop{\smile\hskip-0.9em ^| \ }_C B$.*

PROOF. Let x be a tuple of variables of length α and let $p(x) \in S_\alpha(\emptyset)$. By local character, for each $a \models p$ there is some cardinal $\kappa(a)$ witnessing the property for a. By invariance $\kappa(a)$ is the same for each $a \models p$. Let us call it κ_p. Now the supremum of all κ_p for $p(x) \in S_\alpha(\emptyset)$ satisfies the required condition. ⊣

DEFINITION 12.18. We will be dealing with an arbitrary independence relation $\mathop{\smile\hskip-0.9em ^| \ }$ and we would like to use the standard terminology developed for nonforking independence $\mathop{\smile\!\!\!\!\!\!|}^{f}$ in simple theories. By Corollary 12.6 we know that $\mathop{\smile\!\!\!\!\!\!|}$ is symmetric. Therefore $\mathop{\smile\!\!\!\!\!\!|}$ is also right transitive, that is,

$$\text{if } B \subseteq C \subseteq D,\ A \mathop{\smile\!\!\!\!\!\!|}_B C \text{ and } A \mathop{\smile\!\!\!\!\!\!|}_C D, \text{ then } A \mathop{\smile\!\!\!\!\!\!|}_B D$$

and has right finite character, that is,

$$\text{if } A \mathop{\smile\!\!\!\!\!\!|}_C B_0 \text{ for all finite } B_0 \subseteq B, \text{ then } A \mathop{\smile\!\!\!\!\!\!|}_C B.$$

If I is a linearly ordered set, we say that $(a_i : i \in I)$ is $\mathop{\smile\!\!\!\!\!\!|}$-*independent* over C when $a_i \mathop{\smile\!\!\!\!\!\!|}_C a_{<i}$ for all $i \in I$. Such a sequence will be called a $\mathop{\smile\!\!\!\!\!\!|}$-*Morley sequence over* C if additionally it is C-indiscernible.

Let $A \subseteq B$, $p(x) \in S(A)$ and $p(x) \subseteq q(x) \in S(B)$. We say that $q(x)$ is a $\mathop{\smile\!\!\!\!\!\!|}$-*free extension* of $p(x)$ if for some (all) $a \models q$, $a \mathop{\smile\!\!\!\!\!\!|}_A B$. In this case we also say that q is $\mathop{\smile\!\!\!\!\!\!|}$-free over A.

We say that $\mathop{\smile\!\!\!\!\!\!|}$ satisfies the *Independence Theorem over* C, if whenever

$$a \equiv_C b,\ C \subseteq A \cap B,\ A \mathop{\smile\!\!\!\!\!\!|}_C B,\ a \mathop{\smile\!\!\!\!\!\!|}_C A, \text{ and } b \mathop{\smile\!\!\!\!\!\!|}_C B,$$

then there is some c such that

$$c \mathop{\smile\!\!\!\!\!\!|}_C AB,\ c \equiv_A a, \text{ and } c \equiv_B b.$$

In other terms, if $C \subseteq A \cap B$ and $A \mathop{\smile\!\!\!\!\!\!|}_C B$, for any two types $p(x) \in S(A)$ and $q(x) \in S(B)$ which are $\mathop{\smile\!\!\!\!\!\!|}$-free over C and have a common restriction to C, their union can be extended to a complete type over AB which is $\mathop{\smile\!\!\!\!\!\!|}$-free over C.

PROPOSITION 12.19. $\mathop{\smile\!\!\!\!\!\!|}^{d}$ *is stronger than any independence relation* $\mathop{\smile\!\!\!\!\!\!|}$, *that is: if* $A \mathop{\smile\!\!\!\!\!\!|}^{d}_C B$, *then* $A \mathop{\smile\!\!\!\!\!\!|}_C B$.

PROOF. Assume $a \mathop{\smile\!\!\!\!\!\!|}^{d}_C b$ but $a \not\mathop{\smile\!\!\!\!\!\!|}_C b$. Let $\kappa(a)$ be the cardinal given for a by the local character property and choose a regular cardinal $\kappa > \kappa(a)$. Since $b \mathop{\smile\!\!\!\!\!\!|}_C C$, there is a $\mathop{\smile\!\!\!\!\!\!|}$-Morley sequence $(b_i : i < \kappa)$ over C starting with $b_0 = b$. Its initial segment $(b_i : i < \omega)$ can be obtained as in Proposition 12.4 (using freely the symmetry of $\mathop{\smile\!\!\!\!\!\!|}$) and for its extension to a sequence of length κ we need only to preserve C-indiscernibility since $\mathop{\smile\!\!\!\!\!\!|}$-independence over C is granted by invariance and finite character. Now let $p(x, y) = \mathrm{tp}(ab/C)$. Since $p(x, b)$ does not divide over C, $\bigcup_{i<\kappa} p(x, b_i)$ is consistent. Let a' be a realization of this union of types. Then $a'b_i \equiv_C ab$ for all $i < \kappa$, which implies that $a' \not\mathop{\smile\!\!\!\!\!\!|}_C b_i$ for all $i < \kappa$. If $a' \mathop{\smile\!\!\!\!\!\!|}_{Cb_{<i}} b_i$ then (by transitivity) $a' \mathop{\smile\!\!\!\!\!\!|}_C b_i$, which is not the case. Hence $a' \not\mathop{\smile\!\!\!\!\!\!|}_{Cb_{<i}} b_i$ for all $i < \kappa$. But this contradicts the choice of κ since $\kappa(a) = \kappa(a')$ and therefore $a' \mathop{\smile\!\!\!\!\!\!|}_{Cb_{<i}} (b_j : j < \kappa)$ for some $i < \kappa$. ⊣

LEMMA 12.20. *Let* $\mathop{\smile\!\!\!\!\!\!|}$ *be an independence relation. Assume* $\mathop{\smile\!\!\!\!\!\!|}$ *satisfies the Independence Theorem over* C. *Then for any* $p(x, y) \in S(C)$, *if* $(a_i : i < \alpha)$ *is an* $\mathop{\smile\!\!\!\!\!\!|}$*-independent sequence over* C *and each* $p(x, a_i)$ *is a* $\mathop{\smile\!\!\!\!\!\!|}$*-free extension of its common restriction to* C, *then* $\bigcup_{i<\alpha} p(x, a_i)$ *is* $\mathop{\smile\!\!\!\!\!\!|}$*-free over* C.

PROOF. We inductively construct a chain of types $(q_i : i < \alpha)$ such that $q_i(x) \in S(Ca_{<i})$ extends $\bigcup_{j<i} p(x, a_j)$ and is $\mathop{\smile\!\!\!\!\!\!|}$-free over C. We begin with $q_0 = p(x, a_0) \restriction C$ and for limit i we put $q_i = \bigcup_{j<i} q_j$ (which is $\mathop{\smile\!\!\!\!\!\!|}$-free by the induction hypothesis and finite character). For the case q_{i+1} we apply the Independence Theorem to

$$A = Ca_{<i},\ B = Ca_i,\ q_i(x) \in S(A),\ \text{and}\ p(x, a_i) \in S(B)$$

(which are $\mathop{\smile\!\!\!\!\!\!|}$-free extensions of q_0) obtaining a type $q_{i+1}(x) \in S(AB) = S(Ca_{<i+1})$ extending $p(x, a_i) \cup q_i(x)$ and $\mathop{\smile\!\!\!\!\!\!|}$-free over C. Since $\bigcup_{i<\alpha} q_i(x)$ is $\mathop{\smile\!\!\!\!\!\!|}$-free over C and contains $\bigcup_{i<\alpha} p(x, a_i)$, also $\bigcup_{i<\alpha} p(x, a_i)$ is $\mathop{\smile\!\!\!\!\!\!|}$-free over C. ⊣

THEOREM 12.21. *T is simple if and only if there is an independence relation* $\mathop{\smile\!\!\!\!\!\!|}$ *in T which satisfies the Independence Theorem over models. Moreover if T is simple and* $\mathop{\smile\!\!\!\!\!\!|}$ *is as indicated, then* $\mathop{\smile\!\!\!\!\!\!|} = \mathop{\smile\!\!\!\!\!\!|}^d$.

PROOF. If T is simple then clearly $\mathop{\smile\!\!\!\!\!\!|}^d = \mathop{\smile\!\!\!\!\!\!|}^f$ is an independence relation (see Proposition 12.16) and satisfies the Independence Theorem over models (see Corollary 10.10). For the opposite direction, by Proposition 12.19 we know that $\mathop{\smile\!\!\!\!\!\!|}^d \subseteq \mathop{\smile\!\!\!\!\!\!|}$. We will show now that $\mathop{\smile\!\!\!\!\!\!|} \subseteq \mathop{\smile\!\!\!\!\!\!|}^d$. From this it will follow that $\mathop{\smile\!\!\!\!\!\!|} = \mathop{\smile\!\!\!\!\!\!|}^d$ and hence that $\mathop{\smile\!\!\!\!\!\!|}^d$ has local character in T. By Proposition 12.16 T is simple.

Let $a \mathop{\smile\!\!\!\!\!\!|}_C b$. We check that $a \mathop{\smile\!\!\!\!\!\!|}^d_C b$. Let $p(x, y) = \text{tp}(ab/C)$ and let $(b_i : i < \omega)$ be C-indiscernible with $b_0 = b$. We will show that $\bigcup_{i<\omega} p(x, b_i)$ is consistent. Let $\kappa(b)$ be the cardinal number given for b by the local character property and choose a regular cardinal $\kappa > \kappa(b)$. Extend the given sequence to a C-indiscernible sequence $(b_i : i \leq \kappa)$. By Corollary 1.7 there is a model $M \supseteq C$ such that $(b_i : i \leq \kappa)$ is M-indiscernible. Starting with $M_0 = M$ it is easy now to construct a chain of models $(M_i : i < \kappa)$ such that $Cb_{<i} \subseteq M_i$ and $(b_j : i < j \leq \kappa)$ is M_i-indiscernible. Since $\kappa(b) = \kappa(b_\kappa)$, by choice of κ, $b_\kappa \mathop{\smile\!\!\!\!\!\!|}_{M_i} (M_j : j < \kappa)$ for some $i < \kappa$. Then $b_\kappa \mathop{\smile\!\!\!\!\!\!|}_{M_i} (b_j : i < j < \kappa)$. By invariance and finite character, $(b_j : i < j < \kappa)$ is $\mathop{\smile\!\!\!\!\!\!|}$-independent over M_i and hence it is a $\mathop{\smile\!\!\!\!\!\!|}$-Morley sequence over M_i. By conjugation over C, $(b_i : i < \omega)$ is a $\mathop{\smile\!\!\!\!\!\!|}$-Morley sequence over some model $M \supseteq C$. Let $q(x) \in S(Mb_0)$ be an extension of $p(x, b_0)$ which is $\mathop{\smile\!\!\!\!\!\!|}$-free over C and choose $p'(x, y) \in S(M)$ such that $q(x) = p'(x, b_0)$. Then $p'(x, b_i) \in S(Mb_i)$ is $\mathop{\smile\!\!\!\!\!\!|}$-free over M (in fact over C). By Lemma 12.20, $\bigcup_{i<\omega} p'(x, b_i)$ is consistent. In particular $\bigcup_{i<\omega} p(x, b_i)$ is consistent. ⊣

THEOREM 12.22. *T is stable if and only if there is an independence relation $\mathop{\smile\!\!\!\!\!\!|}$ in T which satisfies one of the two equivalent conditions:*

1. *Types over models are $\mathop{\smile\!\!\!\!\!\!|}$-stationary, that is, for any $p(x) \in S(M)$, for any $B \supseteq M$ there is a unique $\mathop{\smile\!\!\!\!\!\!|}$-free extension of p over B.*
2. *Every type has a bounded number of $\mathop{\smile\!\!\!\!\!\!|}$-free extensions, that is, for each tuple of variables x there is a cardinal μ such that for every $p(x) \in S(A)$ for every $B \supseteq A$ there are at most μ $\mathop{\smile\!\!\!\!\!\!|}$-free extensions of p over B.*

Moreover if T is stable and $\mathop{\smile\!\!\!\!\!\!|}$ is as indicated, then $\mathop{\smile\!\!\!\!\!\!|} = \mathop{\smile\!\!\!\!\!\!|}^d$.

PROOF. If T is stable, T is simple and $\mathop{\smile\!\!\!\!\!\!|}^d = \mathop{\smile\!\!\!\!\!\!|}^f$ is an independence relation. Moreover (see Remark 11.5 and Proposition 11.8) conditions 1 and 2 hold.

1 implies 2. Let α be the length of x and let κ be the cardinal given by local character according to Remark 12.17. Let $\mu = 2^{|T|+\kappa}$. We want to show that μ is an upper bound for the number of $\mathop{\smile\!\!\!\!\!\!|}$-free extensions of $p(x) \in S(A)$ over any other larger set. For this we may assume that $|A| \leq \kappa$ because there is some $C \subseteq A$ of cardinality $< \kappa$ such that p is $\mathop{\smile\!\!\!\!\!\!|}$-free over C and then a bound for $p \restriction C$ is also a bound for p. There is a model $M \supseteq A$ of cardinality κ. The number of extensions of p to a complete type over M is bounded by $|S(M)| \leq 2^{|T|+\kappa} = \mu$. Since every type over M is stationary, the number of $\mathop{\smile\!\!\!\!\!\!|}$-free extensions of p over any set is also bounded by μ.

2 implies stability of T and $\mathop{\smile\!\!\!\!\!\!|} = \mathop{\smile\!\!\!\!\!\!|}^d$ (and hence it implies 1). Fix μ as in 2 and fix an n-tuple of variables x. Choose $\kappa > |T|$ witnessing the local character of $\mathop{\smile\!\!\!\!\!\!|}$ for n as in Remark 12.17. Choose $\lambda \geq \mu$ such that $\lambda = \lambda^{<\kappa}$. We show that T is stable in λ. Let $|A| \leq \lambda$. For each $p(x) \in S(A)$ there is some $C \subseteq A$ such that p is $\mathop{\smile\!\!\!\!\!\!|}$-free over C and $|C| < \kappa$. There are $\leq \lambda^{<\kappa} = \lambda$ such subsets $C \subseteq A$, over each such C there are $\leq 2^{|T|+|C|} \leq \lambda^{<\kappa} = \lambda$ types $q(x) \in S(C)$ and for each $q(x) \in S(C)$ there are at most $\mu \leq \lambda$ $\mathop{\smile\!\!\!\!\!\!|}$-free extensions of q over A. The number of types $p(x) \in S(A)$ is therefore bounded by λ. Thus, T is stable.

By Proposition 12.19 we know that $\mathop{\smile\!\!\!\!\!\!|}^d \subseteq \mathop{\smile\!\!\!\!\!\!|}$. To check $\mathop{\smile\!\!\!\!\!\!|} \subseteq \mathop{\smile\!\!\!\!\!\!|}^d$ assume $p(x) = \mathrm{tp}(a/BC)$ divides over C. Every global extension $\mathfrak{p} \in S(\mathfrak{C})$ of p forks over C and therefore (see Corollary 8.7) has unboundedly many C-conjugates. But if p is $\mathop{\smile\!\!\!\!\!\!|}$-free over C, then over any larger set p has an extension which is $\mathop{\smile\!\!\!\!\!\!|}$-free over C and hence the number of its C-conjugates is bounded by μ. Therefore $a \not\mathop{\smile\!\!\!\!\!\!|}_C B$. ⊣

PROPOSITION 12.23. *The following are equivalent.*

1. *T is not simple.*

2. *For some* $\varphi(x, y) \in L$, *some* $k < \omega$, *there is an indiscernible sequence* $(c_i a_i : i < \omega)$ *such that for all* $i < \omega$, $\models \varphi(c_i, a_0)$ *and* $\varphi(x, a_i)$ *k-divides over* $\{c_j a_j : j < i\}$.
3. *For some* $\varphi(x, y) \in L$, *some* $k < \omega$, *there are a tuple* c *and some* c-*indiscernible sequence* $(a_i : i < \omega)$ *such that for all* $i < \omega$, $\models \varphi(c, a_i)$ *and* $\varphi(x, a_i)$ *k-divides over* $a_{<i}$.
4. *For some* $\varphi(x, y) \in L$ *there are a tuple* c *and some* c-*indiscernible sequence* $(a_i : i \leq \omega)$ *such that* $\models \varphi(c, a_\omega)$ *and* $\varphi(x, a_\omega)$ *divides over* $\{a_i : i < \omega\}$.

PROOF. $1 \Rightarrow 2$. If T is not simple, then (see Proposition 3.11) for some $\varphi(x, y) \in L$, for some $k < \omega$ there is a sequence $(d_i : i < \omega)$ such that $\{\varphi(x, d_i) : i < \omega\}$ is consistent and $\varphi(x, d_i)$ k-divides over $d_{<i}$ for each $i < \omega$. By Proposition 1.6 we may assume $(d_i : i < \omega)$ is indiscernible. We now inductively define $(c_i a_i : i < \omega)$ in such a way that $\varphi(x, a_i)$ k-divides over $\{c_j a_j : j < i\}$ and $\models \varphi(c_i a_0) \wedge \cdots \wedge \varphi(c_i, a_i)$. Indiscernibility can be obtained again by an application of Proposition 1.6.

We start the construction with $a_0 = d_0$, then choosing c_0 such that $\models \varphi(c_0, a_0)$. Since $\varphi(x, d_1)$ k-divides over a_0, there is an a_0-indiscernible sequence $(b_i : i < \omega)$ such that $b_i \equiv_{a_0} d_1$ and $\{\varphi(x, b_i) : i < \omega\}$ is k-inconsistent. By Proposition 1.6 we may assume it is $a_0 c_0$-indiscernible. Set $a_1 = b_0$ and choose c_1 such that $\models \varphi(c_1, a_0) \wedge \varphi(c_1, a_1)$. Then $\varphi(x, a_1)$ k-divides over a_0, c_0. Changing $(d_i : 2 \leq i < \omega)$ by $(d'_i : 2 \leq i < \omega)$ such that $a_0 a_1 (d'_i : 2 \leq i < \omega) \equiv (d_i : i < \omega)$ if necessary, we can continue carrying out the construction.

$2 \Rightarrow 3$. We can take $\omega + 1$ as index set, in which case $\models \varphi(c_\omega, a_i)$ for all $i < \omega$. Then put $c = c_\omega$ and notice that $(a_i : i < \omega)$ is c–indiscernible and that for all $i < \omega$, $\varphi(x, a_i)$ k-divides over $a_{<i}$.

$3 \Rightarrow 4$. Extend the sequence $(a_i : i < \omega)$ to a c-indiscernible sequence $(a_i : i \leq \omega)$.

$4 \Rightarrow 1$. Assume $\varphi(x, a_\omega)$ k-divides over $\{a_i : i < \omega\}$. By indiscernibility for all $i < \omega$, $\varphi(x, a_i)$ k-divides over $a_{<i}$. By c-indiscernibility, $\models \varphi(c, a_i)$ for all $i < \omega$ and therefore $\{\varphi(x, a_i) : i < \omega\}$ is consistent, which contradicts the simplicity of T by Remark 4.11. ⊣

THEOREM 12.24. *The following are equivalent.*

1. *T is simple.*
2. $\mathop{\smile\!\!\!\!\!\!|}^{d}$ *is symmetric.*
3. $\mathop{\smile\!\!\!\!\!\!|}^{f}$ *is symmetric.*
4. $\mathop{\smile\!\!\!\!\!\!|}^{d}$ *is right transitive.*
5. $\mathop{\smile\!\!\!\!\!\!|}^{f}$ *is right transitive.*

PROOF. By Proposition 5.18 and by the fact that in a simple theory $\mathop{\smile\!\!\!\!\!\!|}^{d} = \mathop{\smile\!\!\!\!\!\!|}^{f}$, conditions 2 and 3 follow from 1.

$2 \Rightarrow 4$ and $3 \Rightarrow 5$. Since $\mathop{\smile\!\!\!\!\!\!|}^{d}$ and $\mathop{\smile\!\!\!\!\!\!|}^{f}$ are left transitive, it is clear that symmetry implies they are right transitive.

$4 \Rightarrow 1$ and $5 \Rightarrow 1$. Fix an ordered set of order type $\omega + 2 + \omega^*$, where ω^* is the reverse order of ω, say

$$0 < 1 < \cdots < \omega < \omega + 1 < \cdots < -2 < -1.$$

Assume T is not simple. By Proposition 12.23 and compactness there is some $\varphi(x, y) \in L$ and some k for which there is an indiscernible sequence $(c_i a_i : i \in \omega + 2 + \omega^*)$ such that for each i, $\varphi(x, a_i)$ k-divides over $\{c_j a_j : j < i\}$ and for all $j \leq i$, $\models \varphi(c_i, a_j)$. Let $I = \{c_i : i \in \omega\}$ and let $J = \{c_i : i \in \omega^*\}$. Since $\models \varphi(c_{\omega+1}, a_\omega)$, $c_{\omega+1} \not\mathop{\smile\!\!\!\!\!\!|}^{d}_{I} J a_\omega$. Since $\text{tp}(c_{\omega+1}/IJ)$ is finitely satisfiable in I, $c_{\omega+1} \mathop{\smile\!\!\!\!\!\!|}^{f}_{I} J$. Since $\text{tp}(c_{\omega+1}/a_\omega IJ)$ is finitely satisfiable in J, $c_{\omega+1} \mathop{\smile\!\!\!\!\!\!|}^{f}_{IJ} a_\omega$. This contradicts the right transitivity of $\mathop{\smile\!\!\!\!\!\!|}^{d}$ and $\mathop{\smile\!\!\!\!\!\!|}^{f}$. $\dashv$

Chapter 13

SUPERSIMPLE THEORIES

DEFINITION 13.1. T is *supersimple* if for every finitary $p \in S(A)$ there is a finite $A_0 \subseteq A$ such that p does not fork over A_0. In other words, for any finite tuple a, for any set A, there is a finite $A_0 \subseteq A$ such that $a \perp_{A_0} A$. By Proposition 4.13 this implies T is simple. T is *superstable* if it is stable and supersimple.

DEFINITION 13.2. $\kappa(T)$ is the least cardinal μ such that for each finite tuple a, for each set A there is some $B \subseteq A$ such that $|B| < \mu$ and $a \perp_B A$. If there is no such cardinal μ we set $\kappa(T) = \infty$.

REMARK 13.3. 1. *T is simple if and only if $\kappa(T) < \infty$ if and only if $\kappa(T) \leq |T|^+$.*
2. *T is supersimple if and only if $\kappa(T) = \omega$.*

PROOF. For 1 use Proposition 4.13. ⊣

PROPOSITION 13.4. *The following are equivalent.*

1. *T is supersimple.*
2. *There is no infinite sequence $(\varphi_i(x, a_i) : i < \omega)$ such that $\{\varphi_i(x, a_i) : i < \omega\}$ is consistent and for each $i < \omega$, $\varphi_i(x, a_i)$ divides (forks) over $a_{<i}$.*
3. *There is no infinite increasing chain $(p_i(x) : i < \omega)$ of types $p_i(x) \in S(A_i)$ such that each p_{i+1} is a forking (dividing) extension of p_i.*

PROOF. Similar to the proof of Proposition 4.13. A forking (dividing) chain of types easily gives a forking (dividing) chain of formulas and conversely. If there are no infinite forking chains of formulas, the theory is simple and therefore forking and dividing coincide. ⊣

DEFINITION 13.5. *Lascar ranks* SU and U are ordinal valued (or ∞) and are defined for finitary complete types over sets. SU is defined by:

- $\mathrm{SU}(p) \geq 0$.
- $\mathrm{SU}(p) \geq \alpha + 1$ if and only if there is a forking extension q of p such that $\mathrm{SU}(q) \geq \alpha$.
- $\mathrm{SU}(p) \geq \alpha$ if and only if $\mathrm{SU}(p) \geq \beta$ for all $\beta < \alpha$ in case α is a limit number.

As usual, $\mathrm{SU}(p) = \infty$ if $\mathrm{SU}(p) \geq \alpha$ for all α, and $\mathrm{SU}(p) = \alpha$ if $\mathrm{SU}(p) \geq \alpha$ but $\mathrm{SU}(p) \not\geq \alpha + 1$. U is defined by the same conditions for 0 and for a limit number α. For a successor ordinal the rule is as follows:

- For $p(x) \in S(A)$, $\mathrm{U}(p) \geq \alpha + 1$ if and only if for each cardinal number λ there is a set $B \supseteq A$ and there are at least λ many types $q(x) \in S(B)$ extending p and such that $\mathrm{U}(q) \geq \alpha$.

For a finite tuple a we will use the notation $\mathrm{SU}(a/A) = \mathrm{SU}(\mathrm{tp}(a/A))$ and $\mathrm{U}(a/A) = \mathrm{U}(\mathrm{tp}(a/A))$.

REMARK 13.6. SU *is a* foundation rank, *the foundation rank of finitary complete types over sets with the relation of being a forking extension. In general, if R is a binary relation, the foundation rank of R is the mapping r assigning to every element of the domain of R an ordinal number (or ∞) according to the following rules:*

1. $r(a) \geq 0$.
2. $r(a) \geq \alpha + 1$ *if and only if* $r(b) \geq \alpha$ *for some b such that aRb.*
3. $r(a) \geq \alpha$ *if and only if* $r(a) \geq \beta$ *for all* $\beta < \alpha$ *if α is a limit number.*

By induction on α (and induction on β in the case $\alpha + 1$) one easily sees that

4. *if* $r(a) \geq \alpha$ *and* $\alpha \geq \beta$ *then* $r(a) \geq \beta$,

and therefore if one defines

5. $r(a) = \infty$ *in case* $r(a) \geq \alpha$ *for all* α,
6. $r(a) = \sup\{\alpha : r(a) \geq \alpha\}$ *otherwise,*

it is clear that $r(a) = \alpha$ *if and only if* $r(a) \geq \alpha$ *and* $r(a) \not\geq \alpha + 1$.

Some properties of SU *are better understood if we bear in mind that it is a foundation rank. The following will be helpful:*

7. *If aRb and* $r(a) < \infty$, *then* $r(a) > r(b)$.
8. *If R is transitive, the rank r is connected: if* $r(a) = \alpha < \infty$ *and* $\beta < \alpha$, *then* $r(b) = \beta$ *for some b such that aRb.*
9. *If there is a sequence* $(a_i : i < \omega)$ *such that* $a = a_0$ *and* $a_i R a_{i+1}$ *for all* $i < \omega$, *then* $r(a) = \infty$.
10. *If there is an ordinal number α such that for all a,* $r(a) \geq \alpha$ *implies* $r(a) = \infty$ *then: if* $r(a) = \infty$, *then there is a sequence* $(a_i : i < \omega)$ *such that* $a = a_0$ *and* $a_i R a_{i+1}$ *for all* $i < \omega$.

PROOF. 7 is clear since, by definition, if $r(b) \geq \alpha$ and aRb then $r(a) \geq \alpha+1$. 8 can be proved by induction on α using 7. For 9, prove that for all i, $r(a_i) \geq \alpha$ for any α by induction on α. For 10 note that the hypothesis implies that if $r(a) = \infty$ then $r(b) = \infty$ for some b such that aRb. ⊣

REMARK 13.7. $\mathrm{SU}(p) = 0$ *if and only if p is algebraic if and only if* $\mathrm{U}(p) = 0$.

PROPOSITION 13.8. *Let T be simple and let* $p(x) \subseteq q(x)$ *be complete finitary types.*

1. *If q is a nonforking extension of p, then* $\mathrm{SU}(p) = \mathrm{SU}(q)$.

2. *If* $\mathrm{SU}(p) - \mathrm{SU}(q) < \infty$, *then* q *is a nonforking extension of* p.

PROOF. 1. Clearly $\mathrm{SU}(p) \geq \mathrm{SU}(q)$. We now prove by induction on α that $\mathrm{SU}(p) \geq \alpha$ implies $\mathrm{SU}(q) \geq \alpha$. Consider the case $\mathrm{SU}(p) \geq \alpha + 1$. Let $p(x) \in S(A)$ and $q(x) \in S(B)$. For some $C \supseteq A$ there is a forking extension $p' \in S(C)$ of p such that $\mathrm{SU}(p') \geq \alpha$. Changing C (and p') if necessary, we may assume that there is some $b \models q$ such that $b \models p'$ and $C \mathop{\smile\!\!\!\!\!\!|}_{Ab} B$. Then $b \mathop{\smile\!\!\!\!\!\!|}_{C} B$, and hence $q' = \mathrm{tp}(b/CB)$ is a nonforking extension of p'. By induction hypothesis, $\mathrm{SU}(q') \geq \alpha$. Since q' is a forking extension of q, $\mathrm{SU}(q) \geq \alpha + 1$. Point 2 is clear and corresponds to point 7 of Remark 13.6. ⊣

PROPOSITION 13.9. *If* T *is stable, then* $\mathrm{U} = \mathrm{SU}$.

PROOF. By Corollary 8.7 in a stable theory a global type $\mathfrak{p} \in S(\mathfrak{C})$ forks over A if and only if it has an unbounded orbit in $\mathrm{Aut}(\mathfrak{C}/A)$. By induction on α we prove that $\mathrm{SU}(p) \geq \alpha$ if and only if $\mathrm{U}(p) \geq \alpha$. Consider the case $\alpha + 1$. Assume $p \in S(A)$, $\mathrm{SU}(p) \geq \alpha + 1$ and $q \in S(B)$ is a forking extension of p with $\mathrm{SU}(q) \geq \alpha$. A nonforking extension $\mathfrak{q} \in S(\mathfrak{C})$ of q has unboundedly many A-conjugates. Fix λ and choose a set $C \supseteq B$ such that $\mathfrak{q} \upharpoonright C$ has λ many A-conjugates. By Proposition 13.8 and by the induction hypothesis $\mathrm{U}(\mathfrak{q} \upharpoonright C) \geq \alpha$ and then all its A-conjugates over C also have U-rank $\geq \alpha$. This means that $\mathrm{U}(p) \geq \alpha + 1$. For the other direction, assume $\mathrm{U}(p) \geq \alpha + 1$ and choose $\lambda > \mathrm{Mlt}(p)$, the number of nonforking extensions of p. There is a set $B \supseteq A$ over which p has λ extensions of U-rank $\geq \alpha$. By choice of λ, one of them, say $q \in S(B)$, is a forking extension. By induction hypothesis $\mathrm{SU}(q) \geq \alpha$. Then $\mathrm{SU}(p) \geq \alpha + 1$. ⊣

LEMMA 13.10. *Let* T *be simple.*

1. *There is some ordinal* α *such that* $\mathrm{SU}(p) \geq \alpha$ *implies* $\mathrm{SU}(p) = \infty$.
2. *If* $\mathrm{SU}(p) = \infty$, *there is a forking extension* q *of* p *such that* $\mathrm{SU}(q) = \infty$.

PROOF. 1. Assume for every ordinal α there is a complete type $p_\alpha(x) \in S(A_\alpha)$ such that $\alpha \leq \mathrm{SU}(p_\alpha) < \infty$. Since there is a subset $B \subseteq A_\alpha$ such that $|B| \leq |T|$ and p_α does not fork over B, by Proposition 13.8 we may assume that in fact $|A_\alpha| \leq |T|$. For each α there are boundedly many types $p(x) \in S(A_\alpha)$ and therefore there is an ordinal β_α such that $\mathrm{SU}(p) \leq \beta_\alpha$ if $p(x) \in S(A_\alpha)$ and $\mathrm{SU}(p) < \infty$. Fix an enumeration a_α of A_α. Clearly $\beta_\alpha = \beta_{\alpha'}$ if $\mathrm{tp}(a_\alpha) = \mathrm{tp}(a_{\alpha'})$. Since there are only boundedly many types $\mathrm{tp}(a_\alpha)$ of such tuples a_α, we can obtain an upper bound of all $\mathrm{SU}(p_\alpha)$, which is a contradiction.

2 follows from 1 as shown in points 9, 10 of Remark 13.6. ⊣

PROPOSITION 13.11. *If* T *is simple, the following are equivalent for any finitary* $p \in S(A)$.

1. $\mathrm{SU}(p) = \infty$.
2. *There is a forking chain of types* $(p_n : n < \omega)$ *starting with* $p = p_0$.
3. *Some* $q \in S(B)$ *extending* p *forks over* AB_0 *for any finite subset* $B_0 \subseteq B$.

PROOF. 2 $\Leftrightarrow$ 3: as in the proof of Proposition 4.13. 1 $\Leftrightarrow$ 2 follows from Lemma 13.10 and points 9, 10 of Remark 13.6. $\dashv$

REMARK 13.12. *If $p(x) \in S(M)$ is not definable, then* $\mathrm{U}(p) = \infty$.

PROOF. As explained in the proof of Proposition 7.9 for each cardinal λ there is a model $N \succeq M$ over which there are λ different strong heirs of p. Since again they are all nondefinable, this can be used to show that $\mathrm{U}(p) = \infty$. $\dashv$

PROPOSITION 13.13. 1. *T is supersimple if and only if $\mathrm{SU}(p) < \infty$ for all finitary p.*
2. *T is superstable if and only if $\mathrm{U}(p) < \infty$ for all finitary p.*

PROOF. 1 follows from Proposition 13.11 and Proposition 13.4.
2. If T is superstable, T is stable and by Proposition 13.9 $\mathrm{SU} = \mathrm{U}$. Since T is also supersimple, by 1 $\mathrm{U}(p) < \infty$ for all p. For the other direction, it is enough to show that stability follows from the condition $\mathrm{U}(p) < \infty$ for all p. If T is not stable then there is a nondefinable finitary type $p(x) \in S(M)$ over some model M. Then we apply Remark 13.12. $\dashv$

REMARK 13.14. *If $\mathrm{SU}(p) = \alpha < \infty$, then for any $\beta < \alpha$ there is some $q \supseteq p$ such that $\mathrm{SU}(q) = \beta$.*

PROOF. By point 8 of Remark 13.6. $\dashv$

NOTATION 13.15. *We will denote by $\alpha \oplus \beta$ the* natural sum *of the ordinals α, β. Every ordinal number α can be written uniquely in* Cantor normal form *as $\alpha = \sum_{i=0}^{k} \omega^{\alpha_i} n_i$ where $\alpha_0 > \cdots > \alpha_k$ are ordinals and $n_0, \ldots, n_k$ are natural numbers > 0. If $\beta = \sum_{i=0}^{j} \omega^{\beta_i} m_i$ is also in Cantor normal form, then $\alpha \oplus \beta = \sum_{i=0}^{l} \omega^{\gamma_i} r_i$ where $\gamma_0 > \cdots > \gamma_l$ enumerates $\alpha_0, \ldots, \alpha_k, \beta_0, \ldots, \beta_j$ and*

$$r_i = \begin{cases} n_p & \text{if } \gamma_i = \alpha_p \notin \{\beta_0, \ldots, \beta_j\} \\ m_p & \text{if } \gamma_i = \beta_p \notin \{\alpha_0, \ldots, \alpha_k\} \\ n_p + m_q & \text{if } \gamma_i = \alpha_p = \beta_q. \end{cases}$$

This sum is the least operation $F : On \times On \to On$ which is strictly increasing in both arguments. Clearly, for natural numbers n, m, $n + m = n \oplus m$.

THEOREM 13.16 (Lascar inequalities). *Let T be simple. If $\mathrm{SU}(ab/A) < \infty$, then*

$$\mathrm{SU}(a/Ab) + \mathrm{SU}(b/A) \leq \mathrm{SU}(ab/A) \leq \mathrm{SU}(a/Ab) \oplus \mathrm{SU}(b/A).$$

PROOF. It is easy to see by induction on α that if $\mathrm{SU}(a/A) \geq \alpha$, then $\mathrm{SU}(ab/A) \geq \alpha$. Hence $\mathrm{SU}(ab/A) \geq \mathrm{SU}(a/A)$. From $\mathrm{SU}(ab/A) < \infty$ it follows then that $\mathrm{SU}(a/A) < \infty$ and $\mathrm{SU}(b/A) < \infty$. Then we can freely use Proposition 13.8.

To check the inequality

$$\mathrm{SU}(ab/A) \leq \mathrm{SU}(a/Ab) \oplus \mathrm{SU}(b/A),$$

we prove by induction on α that if $\mathrm{SU}(ab/A) \geq \alpha$, then $\mathrm{SU}(a/Ab) \oplus \mathrm{SU}(b/A) \geq \alpha$. This is clear for $\alpha = 0$ and for limit α. Let us consider the case $\alpha + 1$. Assume $\mathrm{SU}(ab/A) \geq \alpha + 1$. For some $B \supseteq A$ we have $\mathrm{SU}(ab/B) \geq \alpha$ and $ab \not\perp_A B$. Since $ab \not\perp_A B$, either $b \not\perp_A B$ or $a \not\perp_{Ab} B$. Therefore $\mathrm{SU}(b/A) > \mathrm{SU}(b/B)$ or $\mathrm{SU}(a/Ab) > \mathrm{SU}(a/Bb)$. By monotonicity of natural addition of ordinal numbers, $\mathrm{SU}(a/Ab) \oplus \mathrm{SU}(b/A) > \mathrm{SU}(a/Bb) \oplus \mathrm{SU}(b/B)$. By the induction hypothesis $\mathrm{SU}(a/Bb) \oplus \mathrm{SU}(b/B) \geq \alpha$. Hence $\mathrm{SU}(a/Ab) \oplus \mathrm{SU}(b/A) \geq \alpha + 1$.

To check the inequality

$$\mathrm{SU}(a/Ab) + \mathrm{SU}(b/A) \leq \mathrm{SU}(ab/A),$$

we show by induction on α that $\mathrm{SU}(ab/A) \geq \mathrm{SU}(a/Ab) + \alpha$ if $\mathrm{SU}(b/A) \geq \alpha$. The cases $\alpha = 0$ and α limit are straightforward. For the case $\alpha + 1$, assume $\mathrm{SU}(b/A) \geq \alpha + 1$. Then for some $B \supseteq A$, $\mathrm{SU}(b/B) \geq \alpha$ and $b \not\perp_A B$. We may assume that $B \perp_{Ab} a$. By induction hypothesis $\mathrm{SU}(ab/B) \geq \mathrm{SU}(a/Bb) + \alpha$. Since $b \not\perp_A B$, also $ab \not\perp_A B$ and then $\mathrm{SU}(ab/A) > \mathrm{SU}(ab/B)$. Since $a \perp_{Ab} B$ we have $\mathrm{SU}(a/Ab) = \mathrm{SU}(a/Bb)$. Therefore $\mathrm{SU}(ab/A) > \mathrm{SU}(ab/B) \geq \mathrm{SU}(a/Bb) + \alpha = \mathrm{SU}(a/Ab) + \alpha$. We then conclude $\mathrm{SU}(ab/A) \geq \mathrm{SU}(a/Ab) + \alpha + 1$. ⊣

COROLLARY 13.17. *If T is simple and* $\mathrm{SU}(ab/A) < \omega$, *then*

$$\mathrm{SU}(ab/A) = \mathrm{SU}(a/Ab) + \mathrm{SU}(b/A).$$

PROOF. As remarked above, for natural numbers n, m, $n + m = n \oplus m$. ⊣

PROPOSITION 13.18. *Let T be simple.*

1. *If $a \in \mathrm{acl}(Ab)$, then* $\mathrm{SU}(ab/A) = \mathrm{SU}(b/A)$.
2. *If* $\mathrm{acl}(aA) = \mathrm{acl}(bA)$ *then* $\mathrm{SU}(a/A) = \mathrm{SU}(b/A)$.

PROOF. 1. Clearly $\mathrm{SU}(ab/A) \geq S(b/A)$. Moreover it is easy to check by induction on α that $\mathrm{SU}(ab/A) \geq \alpha$ implies $\mathrm{SU}(b/A) \geq \alpha$. 2 follows from 1. ⊣

DEFINITION 13.19. An *abstract rank* is a mapping R assigning an ordinal number or ∞ to all finitary complete types over sets and satisfying the following conditions:

1. If $f \in \mathrm{Aut}(\mathfrak{C})$, then $R(p) = R(p^f)$.
2. If $p \subseteq q$, then $R(p) \geq R(q)$.
3. If $p \in S(A)$ and $A \subseteq B$, then there is some extension $q \in S(B)$ of p such that $R(p) = R(q)$.
4. Let $p \in S(A)$ be such that $R(p) < \infty$. There is a cardinal κ such that for each $B \supseteq A$, p has at most κ extensions $q \in S(B)$ such that $R(p) = R(q)$.

REMARK 13.20. *Let R be an abstract rank. If $p \in S(M)$ is not definable, then $R(p) = \infty$.*

PROOF. Choose α minimal for which there is some nondefinable $p \in S(M)$ over some model M with $R(p) = \alpha$. Let κ be the cardinal given by condition 4 in the definition of rank. As shown in the proof of Proposition 7.9 there is a model $N \succeq M$ over which there are κ^+ different strong heirs of p. All are nondefinable and one of them must have rank $< \alpha$, a contradiction. ⊣

PROPOSITION 13.21. *Let R be an abstract rank.*

1. *Let T be stable, $p \subseteq q$, and $R(p) < \infty$. Then $R(p) = R(q)$ if and only if q is a nonforking extension of p.*
2. *If $R(p) < \infty$ for every complete type p, then T is superstable.*

PROOF. 1. Let $p \in S(A)$, $A \subseteq B$, and $p \subseteq q \in S(B)$. We assume T is stable and $R(p) < \infty$. Fix κ, a bound for the extensions of p of rank $R(p)$. We can find a model $M \supseteq A$ such that all nonforking extensions of p over M are A-conjugate in M and such that each forking extension of p over M has more than κ A–conjugates in M. There is an extension $q' \in S(M)$ of q with $R(q) = R(q')$. Now, if q forks over A then also q' forks and therefore q' has more than κ A-conjugates. By definition of rank $R(p) > R(q')$. Now assume $R(p) > R(q)$ and q does not fork over A. Let $q' \in S(M)$ be a nonforking extension of q and choose $r \in S(M)$, an extension of p of rank $R(p) = R(r)$. As shown above, r does not fork over A. By choice of M, q' and r are A-conjugate. Hence $R(p) = R(r) = R(q') = R(q)$.

2. It suffices to show the stability of T, since we then can use point 1 to easily verify that T is supersimple. If T is unstable then some finitary type $p \in S(M)$ is nondefinable. By Remark 13.20 $R(p) = \infty$. ⊣

PROPOSITION 13.22. *In a stable theory* U *is an abstract rank and it is minimal, that is,* $\mathrm{U}(p) \leq R(p)$ *for any other abstract rank R.*

PROOF. If T is stable, then $\mathrm{U} = \mathrm{SU}$ and by Proposition 13.8 whenever $p \subseteq q$ and $\mathrm{U}(p) < \infty$, q is a nonforking extension of p if and only if $\mathrm{U}(p) = \mathrm{U}(q)$. Since in a stable theory a type has only a bounded number of nonforking extensions, the requirements in the definition of abstract rank are fulfilled. Minimality is easily checked showing by induction on α that if R is a rank and $\mathrm{U}(p) \geq \alpha$, then $R(p) \geq \alpha$. ⊣

COROLLARY 13.23. *T is superstable if and only if there is an abstract rank R such that $R(p) < \infty$ for all p.*

PROOF. If T is superstable, then U is an abstract rank and $\mathrm{U}(p) < \infty$ for all p. The rest follows from Proposition 13.21. ⊣

PROPOSITION 13.24. *Let T be stable and let $p(x) \in S(A)$ be finitary.*

1. *If $\mathrm{U}(p) < \infty$, then for any $B \supseteq A$ there are at most $2^{|T|} + |B|$ extensions $q(x) \in S(B)$ of p.*
2. *If $\mathrm{U}(p) = \infty$ then for any cardinal $\lambda \geq |T| + |A|$ there is a set $B \supseteq A$ such that $|B| \leq \lambda$ and p has at least λ^ω extensions $q(x) \in S(B)$.*

PROOF. If T is stable, then $\mathrm{U} = \mathrm{SU}$. By Proposition 13.11, if $\mathrm{U}(p) < \infty$ then any complete type q over $B \supseteq A$ extending p does not fork over AB_0 for some finite $B_0 \subseteq B$. Since there are only $2^{|T|}$ extensions of p to a complete type $q(x) \in S(AB_0)$ for B_0 finite, and since each such type q has at most $2^{|T|}$ nonforking extensions over B, it is easy to check that $2^{|T|} + |B|$ is a correct upper bound for the number of extensions of p over B. On the other hand, if $\mathrm{U}(p) = \infty$ by Lemma 13.10 p has a forking extension q of U-rank ∞. Let $\mathfrak{q}$ be a global nonforking extension of q. Then $\mathfrak{q}$ forks over A and therefore it has an unbounded orbit in $\mathrm{Aut}(\mathfrak{C}/A)$. Note that every complete type between p and $\mathfrak{q}$ has U-rank ∞. Fix a set $A_1 \supseteq A$ such that $|A_1| \leq \lambda$ and for which there are different types $r_i(x) \in S(A_1)$ for $i < \lambda$ which can be extended to A-conjugates of $\mathfrak{q}$. Note that $\mathrm{U}(r_i) = \infty$. Iterating this procedure we obtain a chain of sets $(A_n : n < \omega)$ of cardinality $|A_n| \leq \lambda$ and a tree of types $(p_s : s \in \lambda^{<\omega})$ such that $p_s \in S(A_n)$ if $s \in \lambda^n$, $p_\emptyset = p$, $p_s \subseteq p_{s'}$ if $s \subseteq s'$, $p_s \neq p_{s'}$ if $s \neq s'$ and $\mathrm{U}(p_s) = \infty$. If we put $p_f = \bigcup_{s \subseteq f} p_s$ for $f \in \lambda^\omega$, we obtain a family $(p_f : f \in \lambda^\omega)$ of λ^ω many complete extensions of p over the set $B = \bigcup_{n<\omega} A_n$ of cardinality $|B| \leq \lambda$. ⊣

THEOREM 13.25. *The following are equivalent*:

1. *T is superstable.*
2. *For all A, for all $n < \omega$, $|S_n(A)| \leq |A| + 2^{|T|}$.*
3. *For all $\lambda \geq 2^{|T|}$, T is λ-stable.*
4. *There is some cardinal μ such that for all $\lambda \geq \mu$, T is λ-stable.*

PROOF. 1 ⇒ 2. There are only $2^{|T|}$ finitary types over $\emptyset$, and by Proposition 13.24 and Proposition 13.13 each finitary $p(x) \in S(\emptyset)$ has at most $2^{|T|} + |A|$ complete extensions over A.

It is clear that 2 ⇒ 3 and 3 ⇒ 4.

4 ⇒ 1. If T is not superstable, then, by Proposition 13.13, there is some $p(x) \in S(A)$ such that $\mathrm{U}(p) = \infty$. Choose $\lambda \geq \mu + |T| + |A|$ such that $\lambda^\omega > \lambda$. By Proposition 13.24 there is a set $B \supseteq A$ of cardinality $\leq \lambda$ such that p has at least λ^ω complete extensions over B. Clearly T is not λ-stable. ⊣

Chapter 14

MORE RANKS

DEFINITION 14.1. D-rank is defined for formulas $\varphi(x) \in L(\mathfrak{C})$ (in finitely many variables x) as follows:

1. $D(\varphi(x)) \geq 0$ if and only if $\varphi(x)$ is consistent.
2. $D(\varphi(x)) \geq \alpha + 1$ if and only if for some $\psi(x, y) \in L$ for all cardinal numbers λ there is an infinite sequence $(a_i : i < \lambda)$ such that
   (a) $\{\psi(x, a_i) : i < \lambda\}$ is k-inconsistent for some $k < \omega$,
   (b) $\models \psi(x, a_i) \rightarrow \varphi(x)$ for each $i < \lambda$, and
   (c) $D(\psi(x, a_i)) \geq \alpha$ for each $i < \lambda$.
3. $D(\varphi(x)) \geq \beta$ if and only if $D(\varphi(x)) > \beta$ for all $\beta < \alpha$ for limit α.

If $\varphi(x)$ is inconsistent, $D(\varphi) = -1$ and otherwise it is the supremum of all α such that $D(\varphi) \geq \alpha$. The definition is extended to arbitrary sets of formulas $\pi(x)$ by

$$D(\pi(x)) = \min\{D(\varphi) : \varphi \text{ is a finite conjunction of formulas in } \pi(x)\}.$$

REMARK 14.2. *If $\varphi(x) \in L(A)$, then $D(\varphi(x)) \geq \alpha + 1$ if and only if $\models \psi(x) \rightarrow \varphi(x)$ and $D(\psi(x)) \geq \alpha$ for some $\psi(x) \in L(\mathfrak{C})$ which divides over A.*

PROPOSITION 14.3. 1. *There is an ordinal α such that for all $\varphi(x) \in L(\mathfrak{C})$, if $D(\varphi) \geq \alpha$, then $D(\varphi) = \infty$.*
2. *If $\varphi(x) \in L(A)$ and $D(\varphi(x)) = \infty$, then $D(\psi(x)) = \infty$ for some $\psi(x)$ such that $\models \psi(x) \rightarrow \varphi(x)$ and $\psi(x)$ divides over A.*
3. *$D(\varphi(x)) = \infty$ if and only if there is a sequence $(\varphi_i(x) : i < \omega)$ of consistent formulas $\varphi_i(x) \in L(A_i)$ such that $\varphi = \varphi_0$, $\models \varphi_{i+1}(x) \rightarrow \varphi_i(x)$ and $\varphi_{i+1}(x)$ divides over $\bigcup_{j \leq i} A_j$.*
4. *T is supersimple if and only if $D(\varphi) < \infty$ for all φ.*

PROOF. 1 is easy, as in Lemma 13.10, 2 follows from 1, and 3 follows from 2. Lastly, 4 follows from 3 and Proposition 13.4. ⊣

LEMMA 14.4. 1. *If $\pi_1(x) \vdash \pi_2(x)$, then $D(\pi_1) \leq D(\pi_2)$.*
2. *$D(\pi) = 0$ if and only if π is algebraic.*
3. *$D(\varphi \vee \psi) = \max\{D(\varphi), D(\psi)\}$.*
4. *If $\pi(x)$ is a partial type over A, there is some $p(x) \in S(A)$ such that $\pi \subseteq p$ and $D(\pi) = D(p)$.*

5. *If $\pi(x)$ is a partial type, there is some finite conjunction $\varphi(x)$ of formulas of $\pi(x)$ such that* $\mathrm{D}(\pi) = \mathrm{D}(\varphi)$.

PROOF. 4 follows from 3. As for 3, it is clear that $\mathrm{D}(\varphi), \mathrm{D}(\psi) \leq \mathrm{D}(\varphi \vee \psi)$. Then it suffices to show that if $\mathrm{D}(\varphi \vee \psi) \geq \alpha$, then $\mathrm{D}(\varphi) \geq \alpha$ or $\mathrm{D}(\psi) \geq \alpha$, and this can be shown by induction on α. Consider the case $\alpha + 1$. Assume $\mathrm{D}(\varphi \vee \psi) \geq \alpha + 1$. For some θ and A, $\models \theta \rightarrow (\varphi \vee \psi)$, $(\varphi \vee \psi) \in L(A)$, θ divides over A, and $\mathrm{D}(\theta) \geq \alpha$. Note that $\models \theta \leftrightarrow (\theta \wedge \varphi) \vee (\theta \wedge \psi)$ and hence the induction hypothesis gives $\mathrm{D}(\theta \wedge \varphi) \geq \alpha$ or $\mathrm{D}(\theta \wedge \psi) \geq \alpha$. We conclude $\mathrm{D}(\varphi) \geq \alpha + 1$ or $\mathrm{D}(\psi) \geq \alpha + 1$. ⊣

REMARK 14.5. *If T is simple, then* $\mathrm{SU} \leq \mathrm{D}$.

PROOF. By induction on α it is easy to check that $\mathrm{SU}(p) \geq \alpha$ implies $\mathrm{D}(p) \geq \alpha$. ⊣

REMARK 14.6. 1. *Let T be simple. If $p(x) \in S(B)$ forks over $A \subseteq B$, and $\mathrm{D}(p) < \infty$, then $\mathrm{D}(p) < \mathrm{D}(p \upharpoonright A)$.*
2. *There is an example of a supersimple theory where there is a type $p(x) \in S(A)$ of $\mathrm{D}(p) = 2$ having a nonforking extension $q(x) \in S(B)$ such that $\mathrm{D}(q) = 1$.*

PROOF. 1. Let $\alpha = \mathrm{D}(p)$, choose $\varphi(x) \in p$ with $\mathrm{D}(\varphi) = \mathrm{D}(p \upharpoonright A)$ and choose $\psi(x) \in p$ which forks over A. Then $\psi'(x) = \psi(x) \wedge \varphi(x) \in p$ and hence $\mathrm{D}(\psi'(x)) \geq \alpha$. Since $\psi'(x)$ forks over A and implies $\varphi(x)$, $\mathrm{D}(p \upharpoonright A) = \mathrm{D}(\varphi(x)) \geq \alpha + 1$. For 2 see Example 5.1.15 in [41]. ⊣

DEFINITION 14.7. The *continuous rank* RC (also denoted R^∞) is defined for all sets of formulas (in finitely many variables) as follows:

1. $\mathrm{RC}(\pi(x)) \geq 0$ if and only if $\pi(x)$ is consistent.
2. $\mathrm{RC}(\pi(x)) \geq \alpha + 1$ if and only if for any conjunction $\varphi(x)$ of formulas in $\pi(x)$ for any cardinal λ there is a sequence $(\pi_i(x) : i < \lambda)$ of partial types $\pi_i(x) \ni \varphi(x)$ such that $\mathrm{RC}(\pi_i) \geq \alpha$ and $\pi_i \cup \pi_j$ is inconsistent for all $i < j < \lambda$.
3. $\mathrm{RC}(\pi(x)) \geq \beta$ if and only if $\mathrm{RC}(\pi(x)) \geq \alpha$ for all $\alpha < \beta$ if β is a limit number.

For a formula $\varphi(x)$ we set $\mathrm{RC}(\varphi) = \mathrm{RC}(\{\varphi\})$.

LEMMA 14.8. 1. *If $\pi(x) \vdash \pi'(x)$, then* $\mathrm{RC}(\pi) \leq \mathrm{RC}(\pi')$.
2. $\mathrm{RC}(\pi) = 0$ *if and only if π is algebraic.*
3. *If $\pi(x)$ is a partial type over A,*

$$\mathrm{RC}(\pi) = \min\{\mathrm{RC}(\varphi) : \varphi \textit{ is a finite conjunction of formulas in } \pi\}$$

and therefore there is a finite conjunction $\varphi(x)$ of formulas in $\pi(x)$ such that $\mathrm{RC}(\pi) = \mathrm{RC}(\varphi)$.
4. $\mathrm{RC}(\pi \cup \{(\varphi \vee \psi)\}) = \max\{\mathrm{RC}(\pi \cup \{\varphi\}), \mathrm{RC}(\pi \cup \{\psi\})\}$.
5. *If $\pi(x)$ is a partial type over A, there is some $p(x) \in S(A)$ such that $\pi \subseteq p$ and* $\mathrm{RC}(\pi) = \mathrm{RC}(p)$.

PROOF. 1. It is an induction on α: if $\mathrm{RC}(\pi) \geq \alpha$, then $\mathrm{RC}(\pi') \geq \alpha$. In the case $\alpha + 1$, given φ, a conjunction of formulas in π', and given a cardinal λ, we first find ψ, a conjunction of formulas in π such that $\psi \vdash \varphi$; next we use the hypothesis $\mathrm{RC}(\pi) \geq \alpha + 1$ to find a sequence $(\pi_i(x) : i < \lambda)$ of pairwise incompatible types $\pi_i \ni \psi$ with $\mathrm{RC}(\pi_i) \geq \alpha$, and then we set $\pi_i' = \pi_i \cup \{\varphi\}$. Since $\pi_i \vdash \pi_i'$, by induction hypothesis $\mathrm{RC}(\pi_i') \geq \alpha$. Hence $(\pi_i' : i < \lambda)$ witnesses that $\mathrm{RC}(\pi') \geq \alpha + 1$.

For 3, choose φ, a conjunction of formulas in π of minimal RC-rank, and show by induction on α that $\mathrm{RC}(\varphi) \geq \alpha$ implies $\mathrm{RC}(\pi) \geq \alpha$.

4. By 1 it is clear that

$$\mathrm{RC}(\pi \cup \{\varphi \vee \psi\}) \geq \max\{\mathrm{RC}(\pi \cup \{\varphi\}), \mathrm{RC}(\pi \cup \{\psi\})\} \geq \alpha.$$

Hence we only have to show that if $\mathrm{RC}(\pi \cup \{\varphi \vee \psi\}) \geq \alpha$, then $\max\{\mathrm{RC}(\pi \cup \{\varphi\}), \mathrm{RC}(\pi \cup \{\psi\})\} \geq \alpha$, and this can be done by induction on α. As usual, we consider only the case $\alpha + 1$. Assume $\mathrm{RC}(\pi \cup \{\varphi\}) \not\geq \alpha + 1$ and $\mathrm{RC}(\pi \cup \{\psi\}) \not\geq \alpha + 1$. Hence we have δ_1, δ_2, conjunctions of formulas in π, and λ_1, λ_2, cardinal numbers, such that there is no sequence $(\pi_i : i < \lambda_1)$ of pairwise incompatible types $\pi_i \ni (\delta_1 \wedge \varphi)$ with $\mathrm{RC}(\pi_i) \geq \alpha$ and there is no sequence $(\pi_i : i < \lambda_2)$ of pairwise incompatible types $\pi_i \ni (\delta_2 \wedge \psi)$ with $\mathrm{RC}(\pi_i) \geq \alpha$. Let $\delta = (\delta_1 \wedge \delta_2)$ and let $\lambda = \max\{\lambda_1, \lambda_2\}$. There is a sequence $(\pi_i : i < \lambda)$ of pairwise incompatible types $\pi_i \ni (\delta \wedge (\varphi \vee \psi))$ with $\mathrm{RC}(\pi_i) \geq \alpha$. Note that $\pi_i \equiv \pi_i \cup \{\delta\} \cup \{\varphi \vee \psi\}$ and then, by 1,

$$\mathrm{RC}(\pi_i \cup \{\delta\} \cup \{\varphi \vee \psi\}) \geq \alpha$$

and by induction hypothesis either $\mathrm{RC}(\pi_i \cup \{\delta\} \cup \{\varphi\}) \geq \alpha$ or $\mathrm{RC}(\pi_i \cup \{\delta\} \cup \{\psi\}) \geq \alpha$. Again by 1, either $\mathrm{RC}(\pi_i \cup \{\delta_1\} \cup \{\varphi\}) \geq \alpha$ or $\mathrm{RC}(\pi_i \cup \{\delta_2\} \cup \{\psi\}) \geq \alpha$. One of these two possibilities takes place λ times, contradicting the choice of λ_1 and λ_2.

5 follows from 4 as in other similar situations. ⊣

REMARK 14.9. $\mathrm{RC}(\pi(x)) \geq \alpha + 1$ *if and only if for each* $\varphi(x)$, *conjunction of formulas of* π, *for each cardinal* λ *there is a set* A *and there is a family* $(p_i(x) : i < \lambda)$ *of different types* $p_i(x) \in S(A)$ *such that* $\mathrm{RC}(p_i) \geq \alpha$ *for all* $i < \lambda$.

PROOF. By point 5 of Lemma 14.8. ⊣

PROPOSITION 14.10. *If* T *is stable, then* $\mathrm{D} = \mathrm{RC}$.

PROOF. It is enough to check it for formulas and then it is clear: after Corollary 8.7, for stable T and $\varphi(x) \in L(A)$, $\mathrm{RC}(\varphi(x)) \geq \alpha + 1$ if and only if there is some $\psi(x)$ such that $\models \psi(x) \to \varphi(x)$, $\psi(x)$ forks over A, and $\mathrm{RC}(\varphi) \geq \alpha$. ⊣

PROPOSITION 14.11. T *is superstable if and only if* $\mathrm{RC}(\varphi) < \infty$ *for every* φ.

PROOF. One direction follows from Proposition 14.10 and point 4 of Proposition 14.3. For the other direction note that $\mathrm{U}(p) \leq \mathrm{RC}(p)$ for any complete type p, and then apply Proposition 13.13. ⊣

DEFINITION 14.12. An abstract rank R is a *continuous rank* if for every α, for every A, $\{p(x) \in S(A) : R(p) < \alpha\}$ is an open subset of $S(A)$.

PROPOSITION 14.13. *If T is stable,* RC *is the smallest continuous rank in T.*

PROOF. By definition and by Lemma 14.8 it is clear that RC always satisfies conditions 1–3 of abstract rank. For condition 4 we need to assume T is stable. By Proposition 14.10 $\mathrm{RC} = \mathrm{D}$. If $p(x) \in S(A)$, $\mathrm{RC}(p) = \alpha < \infty$, and q is a forking extension of p of the same rank $\mathrm{RC}(q) = \alpha$, then q contains a formula $\varphi(x)$ which forks over A. We can assume that $\mathrm{RC}(\varphi) = \alpha$ and that φ implies some $\psi(x) \in p$ of rank $\mathrm{RC}(\psi) = \alpha$. But then $\mathrm{D}(\psi) \geq \alpha + 1$, which is a contradiction. Therefore, all extensions q of p with $\mathrm{RC}(q) = \alpha$ are nonforking extensions and by stability its number is bounded by the multiplicity of p, which is $\leq 2^{|T|}$. It follows that RC is an abstract rank.

Point 3 of Lemma 14.8 implies that RC is continuous. If R is another continuous rank, then by induction on α one sees that if $\mathrm{RC}(p) \geq \alpha$ then $R(p) \geq \alpha$. Consider the case $\alpha + 1$. Let $p(x) \in S(A)$ be such that $\mathrm{RC}(p) \geq \alpha + 1$. We will show that for any $\varphi \in p$ there is some $q \in S(A)$ such that $\varphi \in q$ and $R(q) \geq \alpha + 1$. Continuity of R will then imply $R(p) \geq \alpha + 1$. Now, by Remark 14.9 for each cardinal λ there is some B such that there are at least λ types $q(x) \in S(B)$ such that $\varphi(x) \in q$ and $\mathrm{RC}(q) \geq \alpha$. We may assume that always $A \subseteq B$. Since there are only $2^{|T|+|A|}$ types over A, for some $r(x) \in S(A)$ such that $\varphi \in r$ and for each cardinal λ there is some B such that there are at least λ types $q(x) \in S(B)$ such that $r \subseteq q$ and $\mathrm{RC}(q) \geq \alpha$. By induction hypothesis $R(q) \geq \alpha$ for all such q. By condition 4 in the definition of abstract rank $R(r) \geq \alpha + 1$. ⊣

DEFINITION 14.14. The *Morley rank* of a global type $\mathfrak{p} \in S_n(\mathfrak{C})$, $\mathrm{RM}(\mathfrak{p})$, is its Cantor–Bendixson rank in the space $S_n(\mathfrak{C})$. The Morley rank of a partial type $\pi(x)$, $\mathrm{RM}(\pi)$, (where x is a n-tuple of variables) is the Cantor–Bendixson rank of the closed set $\{\mathfrak{p} \in S_n(\mathfrak{C}) : \pi \subseteq \pi\}$ and its *Morley degree*, $\mathrm{DM}(\pi)$, is the Cantor–Bendixson degree of this closed set. By compactness, $\mathrm{DM}(\pi)$ is finite if $\mathrm{RM}(\pi) < \infty$. It is clear that

$$\mathrm{RM}(\pi) = \max\{\mathrm{RM}(\mathfrak{p}) : \pi \subseteq \mathfrak{p}\}.$$

For a formula φ we set $\mathrm{RM}(\varphi) = \mathrm{RM}(\{\varphi\}$ and $\mathrm{DM}(\varphi) = \mathrm{DM}(\{\varphi\})$.

REMARK 14.15. 1. $\mathrm{RM}(\varphi(x)) \geq 0$ *if and only if* $\varphi(x)$ *is consistent.*
2. $\mathrm{RM}(\varphi(x)) \geq \alpha + 1$ *if and only if there is a sequence* $(\varphi_i(x) : i < \omega)$ *such that* $\models \varphi_i(x) \to \varphi(x)$, $\mathrm{RM}(\varphi_i) \geq \alpha$, *and* $\varphi_i(x) \wedge \varphi_j(x)$ *is inconsistent for all* $i \neq j$.
3. $\mathrm{RM}(\varphi) \geq \alpha$ *if and only if* $\mathrm{RM}(\varphi) \geq \beta$ *for all* $\beta < \alpha$ *if* α *is a limit number.*

PROOF. By Proposition 1.17. ⊣

REMARK 14.16. $\mathrm{RM}(\varphi(x)) \geq \alpha + 1$ *if and only if there for each* $n < \omega$ *there is a sequence* $(\varphi_i(x) : i < n)$ *such that* $\models \varphi_i(x) \to \varphi(x)$, $\mathrm{RM}(\varphi_i) \geq \alpha$, *and* $\varphi_i(x) \wedge \varphi_j(x)$ *is inconsistent for all* $i \neq j$. *Hence the degree* $\mathrm{DM}(\varphi(x))$ *can be defined (in case* $\mathrm{RM}(\varphi) = \alpha < \infty$*) as the maximal* n *for which there is a sequence* $(\varphi_i(x) : i < n)$ *such that* $\models \varphi_i(x) \to \varphi(x)$, $\mathrm{RM}(\varphi_i) \geq \alpha$, *and* $\varphi_i(x) \wedge \varphi_j(x)$ *is inconsistent for all* $i \neq j$.

PROOF. By point 4 of Proposition 1.17. ⊣

PROPOSITION 14.17. *For any partial type* π,

1. $\mathrm{RM}(\pi) = \min\{\mathrm{RM}(\varphi) : \varphi$ *is a conjunction of formulas in* $\pi\}$,
2. $\mathrm{DM}(\pi) = \min\{\mathrm{DM}(\varphi) : \varphi$ *is a conjunction of formulas in* π *and* $\mathrm{RM}(\varphi) = \mathrm{RM}(\pi)\}$.

PROOF. By Proposition 1.16. ⊣

REMARK 14.18. *Morley rank can be computed in any* ω*-saturated model* M *containing the parameters of the type as the Cantor–Bendixson rank in* $S(M)$ *of the closed set determined by the type.*

PROOF. It is enough to check it for formulas and in this case we can use Remark 14.15. The parameters needed in the sequence $(\varphi_i(x) : i < \omega)$ to check that $\mathrm{RM}(\varphi(x)) \geq \alpha + 1$ build a countable sequence and its type over the parameters of φ can be realized in M. ⊣

PROPOSITION 14.19. *Morley rank is a continuous rank.*

PROOF. All conditions in the definition of an abstract rank are easily seen to be satisfied by Morley rank. The bound for the number of extensions with the same rank of a type $p(x)$ is $\mathrm{DM}(p)$. Continuity follows from Proposition 14.17. ⊣

COROLLARY 14.20. *In a stable theory,* $\mathrm{U} \leq \mathrm{RC} \leq \mathrm{RM}$.

PROOF. By propositions 14.19, 13.22, and 14.13. ⊣

DEFINITION 14.21. T is *totally transcendental* if and only if $\mathrm{RM}(\varphi) < \infty$ for all φ.

THEOREM 14.22. 1. *If* T *is* λ*-stable for some* $\lambda < 2^{\omega}$, *then* T *is totally transcendental.*

2. *Any totally transcendental theory is* λ*-stable for all* $\lambda \geq |T|$.

PROOF. 1. Assume $\mathrm{RM}(\varphi) = \infty$. By Remark 14.18, we can work in $S(M)$ for some model M such that $\varphi(x) \in L(M)$ and we can then apply Proposition 1.18, obtaining a tree of formulas $(\varphi_s : s \in 2^{<\omega})$ such that $\varphi_\emptyset = \varphi$, $\mathrm{RM}(\varphi_s) = \infty$, $\varphi_s \equiv (\varphi_{s\frown 0} \vee \varphi_{s\frown 1})$ and $(\varphi_{s\frown 0} \wedge \varphi_{s\frown 1})$ is inconsistent. Every branch $f \in 2^{\omega}$ gives rise to a type $\pi_f = \{\varphi_s : s \subseteq f\}$ and this produces a set of 2^{ω} incompatible partial types over a countable set of parameters, contradicting λ-stability of T if $\lambda < 2^{\omega}$.

2. Let $\lambda \geq |T|$ and let $|A| \leq \lambda$. For each $p(x) \in S(A)$ choose some $\varphi_p(x) \in S(A)$ such that $\mathrm{RM}(p) = \mathrm{RM}(\varphi_p)$ and $\mathrm{DM}(p) = \mathrm{DM}(\varphi_p)$. Since T is totally transcendental, for any $\psi(x) \in L(A)$, $\psi \in p$ if and only if $\mathrm{RM}(\varphi_p \wedge \psi) = \mathrm{RM}(\varphi_p)$ and $\mathrm{DM}(\varphi_p \wedge \psi) = \mathrm{DM}(\varphi_p)$. It follows that $p \neq q$ implies $\varphi_p \neq \varphi_q$. Hence the number $|T| + |A|$ of formulas $\varphi(x) \in L(A)$ is an upper bound for $|S(A)|$. ⊣

COROLLARY 14.23. *Totally transcendental theories, and in particular ω-stable theories, are superstable.*

PROOF. By theorems 14.22 and 13.25. ⊣

DEFINITION 14.24. T is *small* if for all $n < \omega$, $|S_n(\emptyset)| \leq \omega$.

REMARK 14.25. *If T is small, then there is a countable $L_0 \subseteq L$ such that for every $\varphi(x) \in L$ there is some $\varphi'(x) \in L_0$ such that in T, $\varphi(x) \equiv \varphi'(x)$.*

PROOF. Let $S_n(\emptyset) = \{p_i(x) : i < \omega\}$. Choose $\varphi_{i,j}(x) \in p_i$ such that $\neg\varphi_{i,j(x)} \in p_j$ whenever $p_i \neq p_j$. If a, b satisfy the same formulas of $\{\varphi_{i,j} : i, j < \omega\}$ then they have the same type over $\emptyset$ and therefore $\models \varphi(a)$ if and only if $\models \varphi(b)$ for every $\varphi(x) \in L$. Hence, every such $\varphi(x)$ is a boolean combination of the formulas $\varphi_{i,j}(x)$. ⊣

REMARK 14.26. *The following are equivalent*:

1. *T is small.*
2. *For all $n < \omega$, for all finite A, $|S_n(A)| \leq \omega$.*
3. *For all finite A, $|S_1(A)| \leq \omega$.*
4. *T has a saturated countable model.*

PROOF. $1 \Rightarrow 2$ can be justified by a standard counting types argument. $2 \Rightarrow 3$ is clear. For $3 \Rightarrow 4$, the countable saturated model can be constructed as a union $\bigcup_{n\in\omega} A_n$ of countable sets A_n such that each complete 1-type over a finite subset of A_n is realized in A_{n+1}. $4 \Rightarrow 1$ is clear since all $p(x) \in S_n(\emptyset)$ can be realized in the countable saturated model. ⊣

REMARK 14.27. *A theory T is small if and only if the two following conditions hold*:

1. *For all $n < \omega$, for all finite A, the space $S_n(A)$ is scattered, that is, every type in $S_n(A)$ has ordinal Cantor–Bendixson rank.*
2. *There is a countable $L_0 \subseteq L$ such that for every $\varphi(x) \in L$ there is some $\varphi'(x) \in L_0$ such that in T, $\varphi(x) \equiv \varphi'(x)$.*

PROOF. By Remark 14.25 and Corollary 1.19 any small theory satisfies 1 and 2. For the opposite direction, assume 1 and 2 hold in T. By 2, the space $S_n(\emptyset)$ has a countable basis of open sets. By 1 and Proposition 1.20, $S_n(\emptyset)$ is countable. ⊣

REMARK 14.28. 1. *ω-categorical theories are small.*

2. *ω-stable theories are small.*

PROOF. Clear. ⊣

COROLLARY 14.29. *T is ω-stable if and only if T is small, superstable, and every complete type has finite multiplicity.*

PROOF. Let T be ω-stable. By Remark 14.28 T is small and by Corollary 14.23 T is superstable. By Theorem 14.22 T is totally transcendental and therefore the multiplicity of a type is its Morley degree. The other direction is just a counting types argument as in the proof of Theorem 13.25. ⊣

COROLLARY 14.30. *Superstable ω-categorical theories are ω-stable.*

PROOF. By Corollary 14.29, since by ω-categoricity for each finite A there are only finitely many complete n-types over A. Moreover by ω-categoricity if A is finite there is a finest finite A-definable equivalence on relation on n-tuples and, together with the Finite Equivalence Relation Theorem 9.30, this implies that all multiplicities of complete types over A are finite. ⊣

DEFINITION 14.31. An abstract rank R is *cantorian* if and only if any type $p(x) \in S(A)$ has rank $R(p) \geq \alpha + 1$ in the case that p is an accumulation point of $\{q(x) \in S(A) : R(q) \geq \alpha\}$.

PROPOSITION 14.32. *Let $p(x) \in S(A)$. Then* $\mathrm{RM}(p) \geq \alpha + 1$ *if and only if for some $B \supseteq A$ some extension $q(x) \in S(B)$ of p is an accumulation point of* $\{r(x) \in S(B) : \mathrm{RM}(r) \geq \alpha\}$.

PROOF. Let $\mathrm{RM}(p) \geq \alpha + 1$ and choose an ω-saturated model $M \supseteq A$ and let $q(x) \in S(M)$ be an extension of p of Morley rank $\geq \alpha + 1$. By Remark 14.18 q has Cantor–Bendixson rank $\geq \alpha + 1$ in $S(M)$ and therefore it is an accumulation point of types $r(x) \in S(M)$ of Cantor–Bendixson rank $\geq \alpha$. Again by Remark 14.18, these types r have Morley rank $\geq \alpha$.

For the other direction it is enough to prove that $\mathrm{RM}(q) \geq \alpha + 1$, in other words, that RM is cantorian. For this it is enough to show that each $\varphi \in q$ is contained in some $\mathfrak{q} \in S(\mathfrak{C})$ of Cantor–Bendixson rank $\geq \alpha + 1$, that is, φ is contained in infinitely many $\mathfrak{q} \in S(\mathfrak{C})$ of Cantor–Bendixson rank $\geq \alpha$. We know that each such φ is contained in infinitely many types $r(x) \in S(B)$ of Morley rank $\geq \alpha$. But we can choose for each such $r(x) \in S(B)$ an extension $\mathfrak{q}(x) \in S(\mathfrak{C})$ of Cantor–Bendixson rank $\geq \alpha$. ⊣

PROPOSITION 14.33. RM *is the smallest cantorian rank.*

PROOF. By propositions 14.19 and 14.32, RM is a cantorian rank. Let R be another cantorian rank. We prove by induction on α that $\mathrm{RM}(p) \geq \alpha$ implies $R(p) \geq \alpha$. Consider the case $\alpha + 1$. Assume $p(x) \in S(A)$ and $\mathrm{RM}(p) \geq \alpha + 1$. By Proposition 14.32 for some $B \supseteq A$, some $q(x) \in S(B)$ extending p is an accumulation point of $\{r(x) \in S(B) : \mathrm{RM}(r) \geq \alpha\}$. By induction hypothesis, this set is contained in $\{r(x) \in S(B) : R(r) \geq \alpha\}$ and hence q is an accumulation point of this set. Since R is cantorian, $R(q) \geq \alpha+1$ and therefore $R(p) \geq \alpha + 1$. ⊣

THEOREM 14.34. *If T is superstable and ω-categorical, then* U *is cantorian and therefore* $\mathrm{U} = \mathrm{RC} = \mathrm{RM}$.

PROOF. Let $p(x) \in S(A)$ be an accumulation point of $\{q(x) \in S(A) : \mathrm{U}(q) \geq \alpha\}$. By corollaries 14.29 and 14.23, every type has finite multiplicity and hence we can find a finite subset $A_0 \subseteq A$ such that p does not fork over A_0 and $p_0 = p \restriction A_0$ has p as its only nonforking extension over A. By ω-categoricity, the type $p_0(x)$ is isolated by some $\varphi_0(x) \in p_0$. By assumption, there is some $q(x) \in S(A)$ such that $\varphi_0(x) \in q$, $\mathrm{U}(q) \geq \alpha$ and $p \neq q$. It follows that q forks over A_0. Hence $\mathrm{U}(p) = \mathrm{U}(p_0) \geq \mathrm{U}(q) + 1 \geq \alpha + 1$. The rest follows from Proposition 14.33 and Corollary 14.20. ⊣

Chapter 15

HYPERIMAGINARIES

DEFINITION 15.1. For any set A, an *A-hyperimaginary* is an equivalence class $[a]_E$ of a possibly infinite tuple a under a type-definable over A equivalence relation E. We will say that it is a hyperimaginary of *sort* E. In order to simplify notation we set $a_E = [a]_E$ and we often identify the equivalence relation E with the partial type over A which defines E. Without loss of generality we can always assume that the type $E(x, y)$ is closed under conjunction and that all formulas $\varphi(x, y) \in E(x, y)$ are symmetric, that is $\vdash \varphi(x, y) \rightarrow \varphi(y, x)$.

Clearly, A-imaginaries are A-hyperimaginaries. A *hyperimaginary* is a $\emptyset$-hyperimaginary. Thus, real elements and imaginaries are a special case of hyperimaginaries. We sometimes use $\mathfrak{C}^{\text{heq}}$ for the class of all hyperimaginaries. If a is a sequence $a = (a_i : i < \alpha)$ of elements a_i for some ordinal α, we say that α is the *length* of the hyperimaginary a_E. *Finitary hyperimaginaries* are hyperimaginaries of finite length. *Countable hyperimaginaries* are hyperimaginaries of countable length.

DEFINITION 15.2. An automorphism $f \in \text{Aut}(\mathfrak{C})$ fixes a hyperimaginary a_E if $f(a_E) = a_E$, that is, if $\models E(a, f(a))$. Let A be a set of hyperimaginaries. A hyperimaginary e is definable over A if e is fixed by all automorphisms fixing A pointwise. The *definable closure of A*, $\text{dcl}^{\text{heq}}(A)$, is the class of all hyperimaginaries definable over A. Since a hyperimaginary can have any length, $\text{dcl}^{\text{heq}}(A)$ is a proper class. As usual, if a is a sequence of hyperimaginaries, $\text{dcl}^{\text{heq}}(a)$ is defined as $\text{dcl}^{\text{heq}}(A)$ where A is the set enumerated in a. As in the case of real elements, we extend the definition to arbitrary classes A by letting $\text{dcl}^{\text{heq}}(A)$ be the union of $\text{dcl}^{\text{heq}}(B)$ for all subsets $B \subseteq A$.

Notice that if A is a class of imaginaries then $\text{dcl}^{\text{eq}}(A) = \text{dcl}^{\text{heq}}(A) \cap \mathfrak{C}^{\text{eq}}$. We say that the sequences of hyperimaginaries a and b are *equivalent* if $\text{dcl}^{\text{heq}}(a) = \text{dcl}^{\text{heq}}(b)$. This means that a, b are interdefinable, which is clearly equivalent to $\text{Aut}(\mathfrak{C}/a) = \text{Aut}(\mathfrak{C}/b)$. In this situation we write $a \sim b$.

LEMMA 15.3. *Any sequence of hyperimaginaries is equivalent to a hyperimaginary.*

PROOF. Let $a = ([a_i]_{E_i} : i \in I)$ be a sequence of hyperimaginaries, where E_i is an equivalence relation on J_i-sequences and $a_i = (a_{(i,j)} : j \in J_i)$.

Put $K = \bigcup_{i\in I}\{i\} \times J_i$ and consider the equivalence relation E defined for $x = (x_{(i,j)} : (i, j) \in K)$ and $y = (y_{(i,j)} : (i, j) \in K)$ by

$$E(x, y) \leftrightarrow \bigwedge_{i\in I} E_i((x_{(i,j)} : j \in J_i), (y_{(i,j)} : j \in J_i)).$$

Clearly $e = [(a_{(i,j)} : (i, j) \in K)]_E$ is a hyperimaginary and $a \sim e$. ⊣

LEMMA 15.4. *Any hyperimaginary is equivalent to a sequence of countable hyperimaginaries.*

PROOF. Recall that we may assume that the type $E(x, y)$ defining the equivalence relation E is closed under conjunction and all its formulas are symmetric. It will be enough to find for each $\varphi(x, y) \in E(x, y)$ a countable partial type $E_\varphi(x, y) \subseteq E(x, y)$ containing $\varphi(x, y)$ which defines an equivalence relation. Given $\varphi(x, y) \in E(x, y)$ we set $E_\varphi = \{\varphi_n : n \in \omega\}$, where $\varphi_0 = \varphi$ and $\varphi_{n+1}(x, y) \in E(x, y)$ satisfies $\vdash \varphi_{n+1}(x, y) \wedge \varphi_{n+1}(y, z) \rightarrow \varphi_n(x, z)$. The existence of such a φ_{n+1} follows by compactness from the fact that $E(x, y) \cup E(y, z) \vdash \varphi_n(x, z)$. ⊣

LEMMA 15.5. *Let $\pi(x)$ be a partial type over A. If E is an equivalence relation on realizations of π and it is type-definable over A, then there exists an equivalence relation F defined for all sequences of the length of x which is type-definable over A and agrees with E in $\pi(\mathfrak{C})$.*

PROOF. Set $F(x, y) \Leftrightarrow (\pi(x) \wedge \pi(y) \wedge E(x, y)) \vee x = y$. ⊣

PROPOSITION 15.6. *Let e be a hyperimaginary and let b be a real tuple. If $e \in \mathrm{dcl}^{\mathrm{heq}}(b)$, then $e \sim b_E$ for some 0-type-definable equivalence relation E.*

PROOF. Let $e = a_F$. Since a_F is type-definable over a and it is b-invariant, it is type-definable over b and there is a partial type $\pi(x, y)$ over $\emptyset$ such that $\pi(x, b)$ defines a_F. Let $p(y) = \mathrm{tp}(b)$. If $b' \models p$ then $\pi(x, b')$ defines an F-class, and hence either defines e or a class disjoint with it. Thus $\exists x(\pi(x, y) \wedge \pi(x, z))$ defines an equivalence relation $G(y, z)$ in $p(\mathfrak{C})$. By Lemma 15.5 there is an equivalence relation E which is type-definable over $\emptyset$ and agrees with G in $p(\mathfrak{C})$. It is easy to see that $e \sim b_E$. ⊣

COROLLARY 15.7. *If $e \in \mathrm{dcl}^{\mathrm{heq}}(A)$ for some set A of cardinality $\leq \kappa$ then e is equivalent to a hyperimaginary of length $\leq \kappa$.*

PROOF. It follows from Proposition 15.6. ⊣

DEFINITION 15.8. The *algebraic closure* of A, a set of hyperimaginaries, is the class $\mathrm{acl}^{\mathrm{heq}}(A)$ consisting of all hyperimaginaries having finite orbit under the group of all automorphisms fixing A pointwise, that is

$$\mathrm{acl}^{\mathrm{heq}}(A) = \{b \in \mathfrak{C}^{\mathrm{heq}} : |\{f(b) : f \in \mathrm{Aut}(\mathfrak{C}/A)\}| < \omega\}.$$

The *bounded closure of A* is the class $\mathrm{bdd}(A)$ consisting of all hyperimaginaries having a bounded orbit under the group of all automorphisms fixing A

pointwise, that is

$$\mathrm{bdd}(A) = \{b \in \mathfrak{C}^{\mathrm{heq}} : |\{f(b) : f \in \mathrm{Aut}(\mathfrak{C}/A)\}| < |\mathfrak{C}|\}.$$

As usual, if a enumerates A we put $\mathrm{acl}^{\mathrm{heq}}(a) = \mathrm{acl}^{\mathrm{heq}}(A)$ and $\mathrm{bdd}(a) = \mathrm{bdd}(A)$. Moreover we extend the definitions $\mathrm{acl}^{\mathrm{heq}}(A)$ and $\mathrm{bdd}(A)$ to arbitrary classes A of hyperimaginaries as customary: we take the union of all $\mathrm{acl}^{\mathrm{heq}}(B)$ (respectively $\mathrm{bdd}(B)$) where B ranges over subsets of A.

A hyperimaginary b is *A-bounded* if $b \in \mathrm{bdd}(A)$ and it is *bounded* if it is $\emptyset$-bounded. The hyperimaginaries a, b are *interbounded* if $\mathrm{bdd}(a) = \mathrm{bdd}(b)$.

REMARK 15.9. $\mathfrak{C}^{\mathrm{eq}} \cap \mathrm{bdd}(A) = \mathfrak{C}^{\mathrm{eq}} \cap \mathrm{acl}^{\mathrm{heq}}(A) = \mathrm{acl}^{\mathrm{eq}}(A)$ *for any class of imaginaries A.*

PROOF. By compactness, if an imaginary has infinitely many conjugates over a set A it has unboundedly many. ⊣

DEFINITION 15.10. We now define the type $\mathrm{tp}(a_E/b_F)$ of a hyperimaginary a_E over some hyperimaginary b_F. For each formula $\varphi(x, y) \in L$ let $\Phi_\varphi(x, y)$ be the set of formulas

$$\exists x'y'(E(x, x') \wedge F(y, y') \wedge \varphi(x', y')).$$

We define $\mathrm{tp}(a_E/b_F)$ as the union of all partial types $\Phi_\varphi(x, b)$ such that $\models \varphi(a', b')$ for some a', b' such that $E(a, a')$ and $F(b, b')$. It is a partial type over b but the choice of another representative b'' in the F-class of b gives an equivalent partial type over b''. As usual, for hyperimaginaries e, c, d we write $e \equiv_c d$ for $\mathrm{tp}(e/c) = \mathrm{tp}(d/c)$ (after choosing the same representative of c).

PROPOSITION 15.11. *The following are equivalent*:

1. $a_E \equiv_{c_F} b_E$.
2. $ac \equiv a'c'$ *for some* a', c' *such that* $E(a', b)$ *and* $F(c', c)$.
3. $a'c' \equiv b'c''$ *for some* a', c', b', c'' *such that* $E(a', a)$, $F(c', c)$, $E(b', b)$ *and* $F(c'', c)$.
4. *There is some* $f \in \mathrm{Aut}(\mathfrak{C}/c_F)$ *such that* $f(a_E) = b_E$.

PROOF. $4 \Rightarrow 1$, $2 \Rightarrow 3$ and $3 \Rightarrow 4$ are clear. For $1 \Rightarrow 2$, notice that if $\mathrm{tp}(a_E/c_F) = \mathrm{tp}(b_E/c_F)$ and $p(x, y) = \mathrm{tp}(a, c)$, then

$$\pi(x, y) = E(x, b) \cup F(y, c) \cup p(x, y)$$

is consistent. If $\models \pi(a', c')$, then $E(a', b)$, $F(c', c)$ and $ac \equiv a'c'$. ⊣

REMARK 15.12. *Types can also be defined for sequences of hyperimaginaries in an analogous way. Assume* $a = ([a_i]_{E_i} : i \in I)$ *and* $b = ([b_j]_{F_j} : j \in J)$ *are sequences of hyperimaginaries. Let* $x = (x_i : i \in I)$ *and* $y = (y_j : j \in J)$ *where* x_i *is of the length of* a_i *and* y_j *is of the length of* b_j. *For each* $\varphi(x, y) \in L$ *let* $x' = (x_i' : i \in I)$ *and* $y' = (y_j' : j \in J)$ *be new sequences of variables and let* $\Phi_\varphi(x, y)$ *be*

$$\exists(x'_i : i \in I)(y'_j : j \in J)(\bigwedge_{i\in I} E_i(x_i, x'_i) \wedge \bigwedge_{j\in J} F_j(y_j, y'_j) \wedge \varphi(x', y')).$$

The type $\text{tp}(a/b)$ *is defined as the union of all types* $\Phi_\varphi(x, (b_j : j \in J))$ *for all* $\varphi(x, y) \in L$ *for which there are* $(a'_i : i \in I)$ *and* $(b'_j : j \in J)$ *such that* $\models \varphi((a'_i : i \in I), (b'_j : j \in J))$, $E_i(a_i, a'_i)$ *for all* $i \in I$ *and* $F_j(b_j, b'_j)$ *for all* $j \in J$. *Notice that this is the same thing as* $\text{tp}(a^*/b^*)$ *where* a^* *and* b^* *are the hyperimaginaries respectively equivalent to* a *and* b *constructed as in the proof of Lemma* 15.3. *Notice also that a consequence of this definition is that for any sequences of hyperimaginaries* $c = (c_i : i \in I)$, *and* $a = (a_i : i \in I)$,

$$a \equiv_b c \text{ if and only if } a \restriction I_0 \equiv_{b\restriction J_0} c \restriction I_0 \text{ for all finite } I_0 \subseteq I, J_0 \subseteq J.$$

DEFINITION 15.13. If B is a set of hyperimaginaries, then $\text{tp}(a/B)$ can be defined as $\text{tp}(a/b)$ where b is a sequence enumerating B, and $a \equiv_B a'$ can be defined as $a \equiv_b a'$. If A is a class of hyperimaginaries, $a \equiv_A a'$ means $a \equiv_B a'$ for every subset $B \subseteq A$.

DEFINITION 15.14. Let E be a 0-type-definable equivalence relation. A *complete type over a hyperimaginary* e (in the variable x) of sort E is a type of the form $p(x) = \text{tp}(a_E/e)$ where a is a real tuple of the length of x. We say that a_E is a realization of p. Of course, $p(x)$ is a partial type over a representative of e but it is complete in the sense that for any $a_E, b_E \models p(x)$ there is some $f \in \text{Aut}(\mathfrak{C}/e)$ such that $f(a_E) = b_E$. After choosing a representative e' for e, $p(x)$ can be written as a partial type $\pi(x, e')$ over e'. The choice of a different representative e'' gives rise to an equivalent partial type $\pi(x, e'') \equiv \pi(x, e')$ now over e''. We denote by $S_E(e)$ the set of all complete types over e in the sort E. This notation and terminology can also be used for types over sets of hyperimaginaries.

It also makes sense to talk of global types $\mathfrak{p}(x) \in S_E(\mathfrak{C})$ of hyperimaginary sort E. They can be seen as realized types $\text{tp}(a/\mathfrak{C})$ in an elementary extension of the monster model $\mathfrak{C}$ or just as unions of a chains of types $p_i(x) \in S_E(A_i)$ for a family of sets $(A_i : i \in On)$ with $\mathfrak{C} = \bigcup_{i\in On} A_i$.

Sometimes we say that a type $p(x) \in S_E(e)$ is of *hyperimaginary sort* to distinguish it from the case of a type $p(x) \in S(e)$ of *real sort*.

PROPOSITION 15.15. *For any hyperimaginary* e, *the equivalence relation* $x \equiv_e y$ *(on real tuples of a given length) is type-definable over any representative of* e.

PROOF. If $e = a_E$, then $x \equiv_e y \Leftrightarrow \exists u(E(a, u) \wedge xa \equiv yu)$. ⊣

REMARK 15.16. *Note that the proof of Proposition* 15.15 *gives an easy proof of the finitary character of* $\equiv_e$ *for any hyperimaginary* e *since for all real tuples* a, b: $a \equiv_e b$ *if and only if* $a' \equiv_e b'$ *for all corresponding finite subtuples* a' *of* a *and* b' *of* b.

REMARK 15.17. *Let a be a real tuple, let e be a hyperimaginary, and let $p(x) = \mathrm{tp}(a/e)$. For any 0-type-definable equivalence relation E,*

$$\mathrm{tp}(a/a_E e) \equiv p(x) \wedge E(x, a).$$

PROOF. Easy to check. ⊣

PROPOSITION 15.18. *For any set of hyperimaginaries A there are hyperimaginaries a, b such that* $\mathrm{bdd}(A) = \mathrm{dcl}^{\mathrm{heq}}(a)$ *and* $\mathrm{acl}^{\mathrm{heq}}(A) = \mathrm{dcl}^{\mathrm{heq}}(b)$.

PROOF. By Lemma 15.4 $\mathrm{bdd}(A) = \mathrm{dcl}^{\mathrm{heq}}(B)$ if B is the class of all hyperimaginaries in $\mathrm{bdd}(A)$ of length $\leq \omega$. For each $\alpha \leq \omega$ there are at most $2^{|T|}$ many 0-type-definable equivalence relations on α-sequences. For each such equivalence relation E there is an upper bound κ_E for the number of hyperimaginaries e_E in B: there are at most $2^{|T|+|A|}$ possibilities for $p(x) = \mathrm{tp}(e/A)$ and for each such $p(x)$ there are boundedly many $d \models p$ with $d_E \in B$. If κ is the supremum of all these κ_E, it follows that $|B| \leq \kappa + 2^{|T|+|A|}$ and we can choose a sequence c enumerating B. By Lemma 15.3, $c \sim b$ for some hyperimaginary b. Clearly $\mathrm{dcl}^{\mathrm{heq}}(b) = \mathrm{dcl}^{\mathrm{heq}}(B) = \mathrm{bdd}(A)$. The case $\mathrm{acl}^{\mathrm{heq}}(A)$ is similar. ⊣

REMARK 15.19. bdd *is a closure operator on subclasses of $\mathfrak{C}$, that is, for all subclasses A, B of $\mathfrak{C}$:*

1. $A \subseteq \mathrm{bdd}(A)$.
2. *If $A \subseteq B$, then* $\mathrm{bdd}(A) \subseteq \mathrm{bdd}(B)$.
3. $\mathrm{bdd}(\mathrm{bdd}(A)) \subseteq \mathrm{bdd}(A)$.

PROOF. Only 3 needs some checking. We may assume A is a set of hyperimaginaries. By Proposition 15.18 there is some hyperimaginary $e \sim \mathrm{bdd}(A)$, and it suffices to prove that $\mathrm{bdd}(e) \subseteq \mathrm{bdd}(A)$. Note that $e \in \mathrm{bdd}(A)$ and let κ be an upper bound for the number of A-conjugates of e. Assume $a_E \in \mathrm{bdd}(e)$ and let λ be an upper bound for the number of e-conjugates of a_E. It follows that a_E has at most $\lambda \cdot \kappa$ conjugates over A and therefore $a_E \in \mathrm{bdd}(A)$. ⊣

LEMMA 15.20. *For any A-hyperimaginary e, there is some hyperimaginary e' such that* $\mathrm{Aut}(\mathfrak{C}/e') = \{f \in \mathrm{Aut}(\mathfrak{C}/A) : f(e) = e\}$, *that is, $e' \sim Ae$. If e is A-bounded, e' is A-bounded too.*

PROOF. Let $e = b_E$ where E is a type-definable over A equivalence relation. Let a enumerate A, let $p(u) = \mathrm{tp}(a)$, and let $E = E(x, y; a)$. We define

$$F(xz, yu) \Leftrightarrow (z = u \wedge p(u) \wedge E(x, y; z)) \vee xz = yu.$$

It is a 0-type-definable equivalence relation. It is easy to see that $e' = ba_F$ is as required. ⊣

PROPOSITION 15.21. $a \overset{\mathrm{KP}}{\equiv}_A b$ *if and only if* $a \equiv_{\mathrm{bdd}(A)} b$.

PROOF. Consider first the case $A = \emptyset$. By Proposition 15.18 we can assume $\mathrm{bdd}(\emptyset)$ is a single hyperimaginary. The equivalence relation $E(a, b) \Leftrightarrow$

$\mathrm{tp}(a/\mathrm{bdd}(\emptyset)) = \mathrm{tp}(b/\mathrm{bdd}(\emptyset))$ is bounded and by Proposition 15.15 it is type-definable over any representative. Since it is invariant, it is also type-definable over $\emptyset$ and hence $\overset{\mathrm{KP}}{\equiv} \subseteq E$. For the other direction, assume $E(a, b)$. Note that $e = [a]_{\overset{\mathrm{KP}}{\equiv}}$ is a bounded hyperimaginary and thus $e \in \mathrm{bdd}(\emptyset)$. Hence there is some $f \in \mathrm{Aut}(\mathfrak{C}/e)$ such that $f(a) = b$, which implies $a \overset{\mathrm{KP}}{\equiv} b$. The general case cannot be obtained by simply applying the case just proven to $T(A)$ since $\mathrm{bdd}(A)$ is the class of all A-bounded hyperimaginaries while $\mathrm{bdd}(\emptyset)$ computed in $T(A)$ is the class of all A-bounded A-hyperimaginaries. But Lemma 15.20 helps to solve this difficulty. ⊣

REMARK 15.22. *$\overset{\mathrm{KP}}{\equiv}_A$ is a finitary relation: $a \overset{\mathrm{KP}}{\equiv}_A b$ if and only if $a' \overset{\mathrm{KP}}{\equiv}_A b'$ for all corresponding finite subtuples a' of a and b' of b.*

PROOF. By Proposition 15.21, Proposition 15.18, and Remark 15.16. ⊣

LEMMA 15.23. *For any 0-type-definable equivalence relation E, the following are equivalent:*

1. $a_E \in \mathrm{dcl}^{\mathrm{heq}}(M)$.
2. $a_E \in \mathrm{bdd}(M)$.
3. *$E(x, a)$ is finitely satisfiable in M.*

PROOF. Clearly 1 implies 2. We will show $2 \Rightarrow 3$ and $3 \Rightarrow 1$. Assume first that some formula $\varphi(x, a) \in E(x, a)$ is not satisfiable in M. For every cardinal κ we can build a coheir sequence over M, $(a_i : i < \kappa)$, starting with $a_0 = a$. If $i < j < \kappa$, then $\models \neg\varphi(a_i, a_j)$ since otherwise, by indiscernibility, $\models \varphi(a, a_j)$ and hence $\varphi(x, a)$ would be satisfiable in M. Since κ can be arbitrarily large and the elements of the coheir sequence have the same type over M and have different E-classes, $a_E \notin \mathrm{bdd}(M)$.

For $3 \Rightarrow 1$, assume $E(x, a)$ is finitely satisfiable in M and let $f \in \mathrm{Aut}(\mathfrak{C}/M)$ and $\varphi(x, y) \in E(x, y)$. We will show that $\models \varphi(a, f(a))$. This will imply $E(a, f(a))$ and therefore $f(a_E) = a_E$. We may assume that $E(x, y)$ is closed under conjunction and hence $\vdash \psi(x, z) \wedge \psi(z, y) \rightarrow \varphi(x, y)$ for some symmetric $\psi(x, y) \in E(x, y)$. Since $\psi(x, a)$ is satisfiable in M, there is some $c \in M$ such that $\models \psi(c, a)$. Since $a \equiv_M f(a)$, we also have $\models \psi(c, f(a))$. From this it follows that $\models \varphi(a, f(a))$. ⊣

PROPOSITION 15.24. $\mathrm{bdd}(b) = \bigcap_{b \in \mathrm{dcl}^{\mathrm{heq}}(M)} \mathrm{dcl}^{\mathrm{heq}}(M)$.

PROOF. If $a \in \mathrm{bdd}(b)$ and $b \in \mathrm{dcl}^{\mathrm{heq}}(M)$, then clearly $a \in \mathrm{bdd}(M)$ and, by Lemma 15.23, we conclude $a \in \mathrm{dcl}^{\mathrm{heq}}(M)$. For the other direction, let us choose a model M such that $b \in \mathrm{dcl}^{\mathrm{heq}}(M)$ (for instance, a model containing a representative of b) and let us choose a cardinal $\kappa > 2^{|T|+|M|}$. If $a \notin \mathrm{bdd}(b)$, there is a family $(a_i : i < \kappa)$ of different b-conjugates of a starting with $a_0 = a$. By choice of κ there are $i < j < \kappa$ such that $a_i \equiv_M a_j$. Hence for some $a' \neq a$ we have $a \equiv_M a'$, which implies $a \notin \mathrm{dcl}^{\mathrm{heq}}(M)$. ⊣

PROPOSITION 15.25. *Let* $p(x) \in S(A)$.

1. *The restriction* $\overset{\text{Ls}}{\equiv}_A\restriction p$ *of* $\overset{\text{Ls}}{\equiv}_A$ *to* $p(\mathfrak{C})$ *is the finest bounded* A*-invariant equivalence relation on realizations of* p.
2. *The restriction* $\overset{\text{KP}}{\equiv}_A\restriction p$ *of* $\overset{\text{KP}}{\equiv}_A$ *to* $p(\mathfrak{C})$ *is the finest bounded type-definable over* A *equivalence relation on realizations of* p.

PROOF. Since $\overset{\text{Ls}}{\equiv}_A\restriction p$ is bounded and A-invariant, it contains the finest bounded A-invariant equivalence relation E on $p(\mathfrak{C})$. Similarly, $\overset{\text{KP}}{\equiv}_A\restriction p$ contains the finest bounded type-definable over A equivalence relation F on $p(\mathfrak{C})$. On the other hand,

$$E(x,y) \vee (\neg p(x) \vee \neg p(y))$$

is bounded and A-invariant, and therefore it contains $\overset{\text{Ls}}{\equiv}_A$. The corresponding result with respect to $\overset{\text{KP}}{\equiv}_A$ and F is more involved. Assume a, b realize $p(x)$ and $a \overset{\text{KP}}{\equiv}_A b$. We will show that $F(a,b)$. By Proposition 15.21 $a \equiv_{\text{bdd}(A)} b$. Choose with Lemma 15.5 an extension F' of F which is an equivalence relation on all sequences of the length of a, is type-definable over A, and agrees with F on p. Then $e = [a]_F = [a]_{F'}$ is an A-bounded A-hyperimaginary. By Lemma 15.20 there is an A-bounded hyperimaginary e' such that $\text{Aut}(\mathfrak{C}/e') = \{f \in \text{Aut}(\mathfrak{C}/A) : f(e) = e\}$. Since $e' \in \text{bdd}(A)$, $a \equiv_{e'} b$, which implies that $f(a) = b$ for some $f \in \text{Aut}(\mathfrak{C}/A)$ such that $f(e) = e$ and therefore implies that $F(a,b)$. ⊣

PROPOSITION 15.26. *If* E *is a bounded equivalence relation on realizations of* $p(x) \in S(A)$ *and it is type-definable over* A, *then there exists a bounded equivalence relation* F *defined for all tuples of the length of* x *which is type-definable over* A *and agrees with* E *in* $p(\mathfrak{C})$.

PROOF. Since $\overset{\text{KP}}{\equiv}_A\restriction p \subseteq E$, it suffices to set $F(x,y) \Leftrightarrow (p(x) \wedge p(y) \wedge E(x,y)) \vee x \overset{\text{KP}}{\equiv}_A y$. ⊣

PROPOSITION 15.27. *Any* A*-bounded* A*-hyperimaginary is an equivalence class of a bounded type-definable over* A *equivalence relation.*

PROOF. Let E be a type-definable over A equivalence relation and let a_E be an A-bounded A-hyperimaginary. Let $p(x) = \text{tp}(a/A)$ and note that each A-conjugate of a_E is an E-class which is also a union of $\overset{\text{KP}}{\equiv}_A$-classes. Hence if F is defined by

$$F(x,y) \Leftrightarrow \exists z(p(z) \wedge E(x,z) \wedge E(y,z)) \vee x \overset{\text{KP}}{\equiv}_A y,$$

then F is a bounded equivalence relation which is type-definable over A and $a_E = a_F$. ⊣

PROPOSITION 15.28. *Let* e, d *be hyperimaginaries such that* $e \in \text{bdd}(d)$ *and let* A *be the set of all* d*-conjugates of* e. *There is some hyperimaginary* c *such that* $\text{Aut}(\mathfrak{C}/c) = \{f \in \text{Aut}(\mathfrak{C}) : f(A) = A\}$.

PROOF. Let $e = a_E$ and $d = b_G$. We may assume every $\varphi(x, y) \in E(x, y)$ is symmetric. Since $e \in \text{bdd}(d)$, for any such $\varphi(x, y)$ there is a maximal $n_\varphi < \omega$ for which there is a sequence $(a_i : i < n_\varphi)$ such that $de \equiv d[a_i]_E$ for each $i < n_\varphi$ and $\models \neg\varphi(a_i, a_j)$ for all $i < j < n_\varphi$. Fix a witnessing sequence $(a_i^\varphi : i < n_\varphi)$. Let $p(z, x) = \text{tp}(d, e)$, let $r_\varphi(z, x_i)_{i<n_\varphi}$ be the type

$$\bigcup_{i<n_\varphi} p(z, x_i) \cup \{\neg\varphi(x_i, x_j) : i < j < n_\varphi\}$$

and let us define $F(z_1, z_2)$ by

$$\bigwedge_{\varphi(x,y)\in E(x,y)} \exists(x_i : i < n_\varphi)(r_\varphi(z_1, x_i)_{i<n_\varphi} \wedge r_\varphi(z_2, x_i)_{i<n_\varphi}).$$

Note that F is independent of the choice of representatives z_1, z_2 in G-classes.

CLAIM 1. *For any* $f \in \text{Aut}(\mathfrak{C})$, $\models F(b, f(b))$ *if and only if* $f(A) = A$.

PROOF OF CLAIM 1. Assume first $f(A) = A$. Therefore $f^{-1}(A) = A$. Clearly, $\models r_\varphi(b, a_i^\varphi)_{i<n_\varphi}$. Note that $[a_i^\varphi]_E \in A$ and hence $f^{-1}([a_i^\varphi]_E) \in A$ and $f^{-1}([a_i^\varphi]_E) \equiv_d e$, that is, $\models p(b, f^{-1}(a_i^\varphi))$ for all $i < n_\varphi$. It follows that $r_\varphi(f(b), a_i^\varphi)_{i<n_\varphi}$ and thus $F(b, f(b))$. For the other direction, let $e' = [a']_E \in A$ (which means $de \equiv de'$) and assume $F(b, f(b))$. For each $\varphi \in E(x, y)$ choose some $(c_i^\varphi : i < n_\varphi)$ such that $\models r_\varphi(b, c_i^\varphi)_{i<n_\varphi}$ and $\models r_\varphi(f(b), c_i^\varphi)_{i<n_\varphi}$. Then $\models \neg\varphi(c_i^\varphi, c_j^\varphi)$ for $i < j < n_\varphi$ and $d[c_i^\varphi]_E \equiv f(d)[c_i^\varphi]_E \equiv de \equiv de' \equiv f(d)f(e') = f(d)[f(a')]_E$ for all $i < n_\varphi$. By maximality of n_φ, $\models \varphi(f(a'), c_i^\varphi)$ for some $i < n_\varphi$. Therefore $E(f(a'), y) \cup p(b, y)$ is consistent and thus $de \equiv d[f(a')]_E = df(e')$, that is, $f(e') \in A$. This proves $f(A) \subseteq A$. Since also $F(b, f^{-1}(b))$, we obtain the equality $f(A) = A$. ⊣

The claim is obviously true for any other b' such that $[b']_G = d$. From the claim it follows that F is an equivalence relation on realizations of $\text{tp}(d)$. Hence we may assume F is an equivalence relation on all sequences of the length of b. Clearly the hyperimaginary b_F fulfils the requirements. ⊣

Chapter 16

HYPERIMAGINARY FORKING

Definition 16.1. Let A be a class of hyperimaginaries and let I be a set linearly ordered by $<$. The sequence of hyperimaginaries $(e_i : i \in I)$ is *indiscernible over* A or it is A*-indiscernible* if for every $n < \omega$, for every two increasing sequences of indices $i_0 < \dots < i_n$ and $j_0 < \dots < j_n$, $e_{i_0}, \dots, e_{i_n} \equiv_A e_{j_0}, \dots, e_{j_n}$. If A is a set, in practice we may always assume that A is a single hyperimaginary. Note that all the hyperimaginaries e_i are in fact of the same sort and hence we can write $e_i = [a_i]_E$ for a single E.

Lemma 16.2. *Let d be a hyperimaginary.*

1. *Let I, J be linearly ordered infinite sets. If $(e_i : i \in I)$ is a d-indiscernible sequence of hyperimaginaries, then there is a d-indiscernible sequence $(c_i : j \in J)$ such that for every $n < \omega$, for every two increasing sequences of indices $i_0 < \dots < i_n \in I$ and $j_0 < \dots < j_n \in J$, $e_{i_0}, \dots, e_{i_n} \equiv_d c_{j_0}, \dots, c_{j_n}$.*
2. *If $(e_i : i \in I)$ and $(d_i : i \in I)$ are d-indiscernible sequences of hyperimaginaries and $(e_i : i \in I_0) \equiv_d (d_i : i \in I_0)$ for each finite subset $I_0 \subseteq I$, then $f((e_i : i \in I)) = (d_i : i \in I)$ for some $f \in \mathrm{Aut}(\mathfrak{C}/d)$.*

Proof. 1. By compactness. For 2 note that it follows that $(e_i : i \in I) \equiv_d (d_i : i \in I)$ (see Remark 15.12). ⊣

Proposition 16.3. *If $(e_i : i \in I)$ is a sequence of hyperimaginaries indiscernible over the hyperimaginary d, then for some representative $\hat{d}$ of d, some sequence $(\hat{e}_i : i \in I)$ of corresponding representatives of $(e_i : i \in I)$ is $\hat{d}$-indiscernible.*

Proof. Fix d', a representative of d. Since the sequence we seek is just a realization of some partial type over d' and representatives of the hyperimaginaries e_i, we may assume $(I, <) = (\omega, <)$. Let κ be an infinite cardinal number larger than the length of d', and larger than the length of every representative of e_i, and let $\lambda = \beth_{(2^\kappa)^+}$. By Lemma 16.2 we can extend $(e_i : i < \omega)$ to a d-indiscernible sequence $(e_i : i < \lambda)$. Choose corresponding representatives $[a_i]_E = e_i$. By Proposition 1.6 there is a d'-indiscernible sequence $(c_i : i < \omega)$ such that for all $n < \omega$ there are some $i_0 < \dots < i_n < \lambda$ such that $c_0 \dots c_n \equiv_{d'} a_{i_0} \dots a_{i_n}$. Since $([c_i]_E : i < \omega) \equiv_d (e_i : i < \omega)$, for some representative $\hat{d}$ of d there exists a $\hat{d}$-indiscernible sequence $(\hat{e}_i : i < \omega)$ such that $[\hat{e}_i]_E = e_i$. ⊣

PROPOSITION 16.4. *Let d be a hyperimaginary.*

1. *For every hyperimaginary $e \notin \mathrm{bdd}(d)$, there is a d-indiscernible sequence $(e_i : i < \omega)$ of distinct hyperimaginaries starting with $e_0 = e$.*
2. *If the sequence of hyperimaginaries $(e_i : i \in I)$ is d-indiscernible, then it is also indiscernible over* $\mathrm{bdd}(d)$.

PROOF. 1. Let $e = a_E$. Since $e \notin \mathrm{bdd}(d)$, by Erdős–Rado (see the proof of Remark 9.2), for some $\varphi(x, y) \in E(x, y)$ there are $(a_i : i < \omega)$ such that $a = a_0 \equiv_d a_i$ and $\models \neg\varphi(a_i, a_j)$ for all $i < j < \omega$. By ordinary methods, like Ramsey's Theorem and compactness, we find a d-indiscernible sequence $(b_i : i < \omega)$ such that $\models \neg\varphi(b_i, b_j)$ for all $i < j < \omega$ and $b_0 \equiv_d a$. Then a d-conjugate of $([b_i]_E : i < \omega)$ satisfies the requirements.

2. By Proposition 16.3 some sequence $(\hat{e}_i : i \in I)$ of representatives is indiscernible over some representative $\hat{d}$ of d. By Corollary 1.7 $(\hat{e}_i : i \in I)$ is indiscernible over some model M containing $\hat{d}$. By Proposition 15.24, $\mathrm{bdd}(d) \subseteq \mathrm{dcl}^{\mathrm{heq}}(M)$ and hence $(\hat{e}_i : i \in I)$ and $(e_i : i \in I)$ are indiscernible over $\mathrm{bdd}(d)$. ⊣

DEFINITION 16.5. The formula $\varphi(x, a)$ *divides* over the hyperimaginary e (with respect to k) if there is some e-indiscernible sequence $(a_i : i < \omega)$ with $a_0 = a$ for which $\{\varphi(x, a_i) : i < \omega\}$ is inconsistent (k-inconsistent). The formula $\varphi(x, a)$ *forks* over e if there are $\psi_1(x, b_1), \dots, \psi_n(x, b_n)$ such that $\varphi(x, a) \vdash \psi_1(x, b_1) \vee \dots \vee \psi_n(x, b_n)$ and each $\psi_i(x, b_i)$ divides over e. The set of formulas $\pi(x)$ *divides* (*forks*) over e if $\pi(x)$ implies some formula which divides (forks) over e. The hyperimaginary a is *independent* of the hyperimaginary b over the hyperimaginary e (written $a \mathop{\smile\!\!\!\!\!\!|}_e b$) if $\mathrm{tp}(a/be)$ (as a partial type over representatives $\hat{b}, \hat{e}$ of b, e) does not fork over e. Note that this is independent of the choice of representatives $\hat{b}, \hat{e}$ because equivalent partial types can be interchanged for dividing and forking purposes. Other notions can be defined in a similar way and we will make use of them when necessary. For instance, a sequence of hyperimaginaries $(e_i : i \in I)$ is a *Morley sequence* over the hyperimaginary e if it is e-indiscernible and $e_i \mathop{\smile\!\!\!\!\!\!|}_e e_{<i}$ for all $i \in I$. We will assume all this also applies to any set A of hyperimaginaries instead of a single hyperimaginary e (replace A by a hyperimaginary e equivalent to a sequence enumerating A).

REMARK 16.6. *For all hyperimaginaries a, b, e: $a \mathop{\smile\!\!\!\!\!\!|}_e b$ if and only if $a \mathop{\smile\!\!\!\!\!\!|}_e eb$.*

PROOF. Obvious, by definition of independence. ⊣

PROPOSITION 16.7. *A partial type $\pi(x)$ divides over the hyperimaginary e with respect to k if and only if it divides over some representative of e with respect to k.*

PROOF. By Proposition 16.3. ⊣

REMARK 16.8. *Let $e \sim e'$ be equivalent hyperimaginaries. A partial type divides over e if and only if it divides over e'. The same is true for forking.*

PROOF. Obviously, it is enough to check it for dividing; this is clear, since a sequence is e-indiscernible if and only if it is e'-indiscernible. ⊣

REMARK 16.9. *Let a, b, c be hyperimaginaries and assume $b \sim b'$ and $c \sim c'$. If $a \mathop{\smile\!\!\!\!\!|}_b c$, then $a \mathop{\smile\!\!\!\!\!|}_{b'} c'$.*

PROOF. By Remark 16.8, since $\text{tp}(a/bc)$ and $\text{tp}(a/b'c')$ are equivalent types. ⊣

REMARK 16.10. *Let $\pi(x, y)$ be a partial type over $\emptyset$. Then $\pi(x, b)$ divides over the hyperimaginary e if and only if for some e-indiscernible sequence $(b_i : i < \omega)$ with $b = b_0$, $\bigcup_{i<\omega} \pi(x, b_i)$ is inconsistent.*

PROOF. By Proposition 16.7, Proposition 16.3, and Remark 4.2. ⊣

LEMMA 16.11. *For all hyperimaginaries a, b, e: $\text{tp}(a/be)$ does not divide over e if and only if for every e-indiscernible sequence $I \ni b$ there is some $a' \equiv_{eb} a$ such that I is ea'-indiscernible.*

PROOF. We adapt the proof of Lemma 4.7. From right to left it is the same as the proof of Lemma 4.7. For the other direction, assume $\text{tp}(a/be)$ does not divide over e and, to simplify notation, let $I = ([b_i]_E : i < \omega)$ be e-indiscernible with $b = [b_0]_E$. By Proposition 16.3 we may assume that $(b_i : i < \omega)$ is indiscernible over some representative $\hat{e}$ of e. Let $\pi(x, \hat{e}, b_0) = \text{tp}(a/eb)$ and let $\Gamma(x, \hat{e}, b_i)_{i<\omega}$ be the set of formulas expressing that $(b_i : i < \omega)$ is indiscernible over $\hat{e}x$. It is enough to show that $\pi(x, \hat{e}, b_0) \cup \Gamma(x, \hat{e}, b_i)_{i<\omega}$ is consistent. By the above remark, $\bigcup_{i<\omega} \pi(x, \hat{e}, b_i)$ is consistent and can be realized by some c. Let $\Gamma_0(x, \hat{e}, b_i)_{i<\omega}$ be a finite subset of $\Gamma(x, \hat{e}, b_i)_{i<\omega}$. By Ramsey's Theorem, there is a one-to-one mapping $f : \omega \to \omega$ such that $\models \Gamma_0(c, \hat{e}, f(b_i))_{i<\omega}$. Now take some c' such that $c'(b_i : i < \omega) \equiv_{\hat{e}} c(f(b_i) : i < \omega)$ and note that c' realizes $\Gamma_0(x, \hat{e}, b_i)_{i<\omega}$ and $\pi(x, \hat{e}, b_0)$. ⊣

PROPOSITION 16.12. *For all hyperimaginaries a, b, c, d: if $\text{tp}(b/cd)$ does not divide over d and $\text{tp}(a/cbd)$ does not divide over bd, then $\text{tp}(ab/cd)$ does not divide over d.*

PROOF. By Lemma 16.11. ⊣

PROPOSITION 16.13.

1. *Let $\pi(x)$ be a partial type over A. If π does not fork over the hyperimaginary e, then some completion $p(x) \in S(A)$ of π does not fork over e.*
2. *Let a, b, c be hyperimaginaries such that $a \mathop{\smile\!\!\!\!\!|}_b c$. Then for every hyperimaginary d, there is some $a' \equiv_{bc} a$ such that $a' \mathop{\smile\!\!\!\!\!|}_b cd$.*

PROOF. 1. As in Remark 4.4, the reason is that $\pi(x) \cup \{\neg\varphi(x) : \varphi(x) \in L(A)$ forks over $e\}$ is consistent.

2. Fix representatives $\hat{b}, \hat{c}, \hat{d}$ of b, c, d and put $\mathrm{tp}(a/bc) = \pi(x, \hat{b}, \hat{c})$. By 1 there exists a completion $p(x) \in S(\hat{b}\hat{c}\hat{d})$ of $\pi(x, \hat{b}, \hat{c})$ which does not fork over b. Let a be of sort E, let $\hat{a} \models p$ and let $a' = \hat{a}_E$. Then $a' \equiv_{bc} a$. Since $\mathrm{tp}(a'/bcd)$, as a partial type over $\hat{b}\hat{c}\hat{d}$, is contained in p, $a' \perp_b cd$. ⊣

PROPOSITION 16.14. *Let e be a hyperimaginary.*

1. *If $\pi(x)$ divides over e and $\pi(x)$ is a partial type over A, then $\pi(x)$ divides over $\hat{e}$ for any representative $\hat{e}$ of e such that $\hat{e} \perp_e A$.*
2. *If T is simple, then $a \perp_e e$ for any hyperimaginary a.*
3. *If T is simple, then a partial type $\pi(x)$ forks over e if and only if $\pi(x)$ forks over some representative of e.*

PROOF. 1. Fix $\varphi(x, y) \in L$ and $b \in A$ such that $\pi(x) \vdash \varphi(x, b)$ and $\varphi(x, b)$ divides over e. Then for some e-indiscernible sequence $(b_i : i < \omega)$ with $b = b_0$, $\{\varphi(x, b_i) : i < \omega\}$ is inconsistent. Since $\hat{e} \perp_e b$, by Lemma 16.11 there is another $\hat{e}$-indiscernible sequence $(b'_i : i < \omega)$ with $b = b'_0$ and such that $\{\varphi(x, b'_i) : i < \omega\}$ is inconsistent. Then $\varphi(x, b)$ divides over $\hat{e}$.

2. Choose a representative $\hat{e}$ of e. We will check that the partial type $\pi(x, \hat{e}) = \mathrm{tp}(a/e)$ does not fork over e. Assume $\pi(x, \hat{e}) \vdash \varphi_1(x, a_1) \vee \cdots \vee \varphi_n(x, a_n)$ where every $\varphi_i(x, a_i)$ divides over e with respect to k_i. Let $k = \max\{k_1, \ldots, k_n\}$, let $\Delta = \{\varphi_1(x, y_1), \ldots, \varphi_n(x, y_n)\}$, and let $m = \mathrm{D}(\pi(x, \hat{e}), \Delta, k)$. By Proposition 3.12 there is a completion $p(x) \in S(\hat{e}a_1, \ldots, a_n)$ of $\pi(x, \hat{e})$ with $\mathrm{D}(p(x), \Delta, k) = m$. For some i, $\varphi_i(x, a_i) \in p$. Now, $\varphi_i(x, a_i)$ divides over e with respect to k and by Proposition 16.7 it divides over some representative $\tilde{e}$ of e with respect to k. Notice that $\pi(x, \hat{e}) \equiv \pi(x, \tilde{e})$. Then $\mathrm{D}(\pi(x, \tilde{e}) \cup \{\varphi_i(x, a_i)\}, \Delta, k) \geq \mathrm{D}(p(x), \Delta, k) = m$ and hence $\mathrm{D}(\pi(x, \hat{e}), \Delta, k) = \mathrm{D}(\pi(x, \tilde{e}), \Delta, k) \geq m + 1$, a contradiction.

3. Assume $\pi(x) \vdash \varphi_1(x, a_1) \vee \cdots \vee \varphi_n(x, a_n)$ where every $\varphi_i(x, a_i)$ divides over e. By 2 and Proposition 16.13 we can choose a representative $\hat{e}$ of e such that $\hat{e} \perp_e a_1, \ldots, a_n$. By 1 every $\varphi_i(x, a_i)$ divides over $\hat{e}$. Hence $\pi(x)$ forks over $\hat{e}$. ⊣

COROLLARY 16.15. *If T is simple, a partial type forks over a hyperimaginary e if and only if it divides over e.*

PROOF. By Proposition 16.14 and Proposition 5.17. ⊣

PROPOSITION 16.16. *Let T be simple. For all hyperimaginaries a, b, c, d:*

1. *If $ab \perp_c d$, then $a \perp_c d$ and $a \perp_{bc} d$.*
2. *If $a \perp_b cd$, then $a \perp_b d$ and $a \perp_{bc} d$.*
3. *There is some $a' \equiv_b a$ such that $a' \perp_b c$.*
4. *If $a \perp_c d$ and $b \perp_{ac} d$, then $ab \perp_c d$.*
5. $a \perp_b \mathrm{bdd}(b)$.

6. *If* $d \perp_{c} ab$, *then* $a \perp_{c} b$ *if and only if* $a \perp_{cd} b$.

PROOF. 1 and 2 follow straightforwardly from Corollary 16.15 and the definition of dividing. To check that $a \perp_{bc} d$ in 1 and 2 it may be convenient to use Remark 16.10 applied to $\mathrm{tp}(a/bcd)$.

3 follows from point 2 in Proposition 16.14 and Proposition 16.13, while 4 follows from Proposition 16.12 and Corollary 16.15.

5. By 3 there is some $a' \equiv_b a$ such that $a' \perp_{b} \mathrm{bdd}(b)$. There is some $f \in \mathrm{Aut}(\mathfrak{C}/b)$ such that $f(a') = a$. Since f fixes setwise $\mathrm{bdd}(b)$, if we apply f we obtain $a \perp_{b} \mathrm{bdd}(b)$.

6. Assume $d \perp_{c} ab$ and $a \perp_{c} b$. By 2 $d \perp_{ac} b$ and then by 4 $da \perp_{c} b$. By 1 $a \perp_{cd} b$. Assume now $d \perp_{c} ab$ and $a \perp_{cd} b$. By 2 $d \perp_{c} b$ and by 4, $ad \perp_{c} b$. Then by 1 $a \perp_{c} b$. ⊣

LEMMA 16.17. *Let T be simple. For any hyperimaginaries a, b, c the following are equivalent*:

1. $a \perp_{b} c$.
2. *There are representatives $\hat{a}, \hat{c}$ of a, c such that* $\hat{a} \perp_{b} \hat{c}$.
3. $a \perp_{\hat{b}} c$ *for every representative $\hat{b}$ of b such that* $\hat{b} \perp_{b} ac$.
4. $a \perp_{\hat{b}} c$ *for some representative $\hat{b}$ of b such that* $\hat{b} \perp_{b} ac$.

PROOF. $1 \Rightarrow 2$. We use the different points of Proposition 16.16 several times. By point 3 of 16.16 there is some representative $\hat{a}$ of a such that $\hat{a} \perp_{a} bc$. By points 4 and 1 of 16.16, $\hat{a} \perp_{b} c$. Let $\hat{b}, \hat{c}$ be representatives of b, c and write $\mathrm{tp}(\hat{a}/bc) = \pi(x, \hat{b}, \hat{c})$. This partial type does not fork over b and by Proposition 16.13 it can be extended to some complete type $p(x) \in S(\hat{b}\hat{c})$ which does not fork over b. Let $a_1 \models p$. Then $a_1 \equiv_{bc} \hat{a}$. Since $p \vdash \mathrm{tp}(a_1/\hat{b}\hat{c}b)$, $a_1 \perp_{b} \hat{b}\hat{c}$ and, in particular, $a_1 \perp_{b} \hat{c}$. Let c_1 be such that $a_1\hat{c} \equiv_{bc} \hat{a}c_1$. Then $\hat{a} \perp_{b} c_1$ and c_1 is also a representative of c.

$2 \Rightarrow 1$. Clear since $\mathrm{tp}(\hat{a}/b\hat{c})$ (as a partial type over the real tuple $\hat{c}$ and a representative $\hat{b}$ of b) includes $\mathrm{tp}(a/bc)$ (as a partial type over $\hat{c}, \hat{b}$).

$1 \Rightarrow 3$. Assume $a \perp_{b} c$ and $\hat{b} \perp_{b} ac$. By point 6 of Proposition 16.16, $a \perp_{b\hat{b}} c$. Since $b\hat{b} \sim \hat{b}$, by Remark 16.9, $a \perp_{\hat{b}} c$.

$3 \Rightarrow 4$. Clear, since there are such $\hat{b}$.

$4 \Rightarrow 1$. Assume $a \perp_{\hat{b}} c$ and $\hat{b} \perp_{b} ac$. By Remark 16.9 $a \perp_{b\hat{b}} c$ and by point 6 of 16.16, $a \perp_{b} c$. ⊣

PROPOSITION 16.18 (Symmetry and transitivity). *If T is simple, then independence is symmetric and transitive for hyperimaginaries, that is, for any hyperimaginaries a, b, c, d*:

1. $a \perp_{b} c$ *if and only if* $c \perp_{b} a$.
2. *If* $a \perp_{b} c$ *and* $a \perp_{bc} d$, *then* $a \perp_{b} cd$.

PROOF. 2 is a consequence of 1 and of point 4 of Proposition 16.16. For 1, use Lemma 16.17 and symmetry of independence for real tuples. ⊣

PROPOSITION 16.19. *Let T be simple. For all hyperimaginaries a, b, c:*

1. $a \mathop{\smile\!\!\!\!\!\mid}_b c$ *if and only if* $a \mathop{\smile\!\!\!\!\!\mid}_b \mathrm{bdd}(c)$ *if and only if* $a \mathop{\smile\!\!\!\!\!\mid}_{\mathrm{bdd}(b)} c$ *if and only if* $\mathrm{bdd}(a) \mathop{\smile\!\!\!\!\!\mid}_b c$.
2. $a \mathop{\smile\!\!\!\!\!\mid}_b a$ *if and only if* $a \in \mathrm{bdd}(b)$.

PROOF. 1 follows easily from point 5 of Proposition 16.16 and the basic properties of independence. Similarly for the direction from right to left of 2. For the other direction, assume $a \notin \mathrm{bdd}(b)$. By Proposition 16.4 there is a b-indiscernible sequence $(a_i : i < \omega)$ with $a_0 = a$ and $a_i \neq a_j$ for all $i < j < \omega$. Let a be of sort E and choose with Proposition 16.3 a b-indiscernible sequence $(\hat{a}_i : i < \omega)$ of representatives. Since $\bigcup_{i<\omega} E(x, \hat{a}_i)$ is inconsistent, the partial type $E(x, \hat{a}_0)$ divides and forks over b. Hence $a \not\mathop{\smile\!\!\!\!\!\mid}_b a$. ⊣

PROPOSITION 16.20 (Finite character). *Let T be simple. Let a, b hyperimaginaries and let $c = (c_i : i \in I)$ a sequence of hyperimaginaries. If for each finite $I_0 \subseteq I$, $a \mathop{\smile\!\!\!\!\!\mid}_b (c_i : i \in I_0)$, then $a \mathop{\smile\!\!\!\!\!\mid}_b c$.*

PROOF. Let $\hat{b}$ be a representative of b and let $(\hat{c}_i : i \in I)$ be a corresponding sequence of representatives for $(c_i : i \in I)$. Write $\mathrm{tp}(a/b(c_i : i \in I))$ as a partial type $\pi(x, \hat{b}, (\hat{c}_i : i \in I))$. For each finite subset Σ of this partial type we can find some finite $I_0 \subseteq I$ such that $\mathrm{tp}(a/b(c_i : i \in I_0))$ can be written as a partial type over $\hat{b}(\hat{c}_i : i \in I_0)$ containing Σ. Therefore Σ does not fork over b. Then $\pi(x, \hat{b}, (\hat{c}_i : i \in I))$ does not fork over b and $a \mathop{\smile\!\!\!\!\!\mid}_b c$. ⊣

PROPOSITION 16.21 (Local character). *Let T be simple, let a be a hyperimaginary of length λ and let $b = (b_i : i \in I)$ be a sequence of hyperimaginaries. Then $a \mathop{\smile\!\!\!\!\!\mid}_{(b_i : i \in J)} b$ for some $J \subseteq I$ such that $|J| \leq |T| + \lambda$.*

PROOF. We may assume $I = \kappa$ for some cardinal κ. Choose inductively representatives $\hat{b}_i$ of b_i such that $\hat{b}_i \mathop{\smile\!\!\!\!\!\mid}_{b_i} b\hat{b}_{<i}$ for all $i < \kappa$. It is then easy to see that for all subsets J of κ, $(\hat{b}_i : i \in J) \mathop{\smile\!\!\!\!\!\mid}_{(b_i : i \in J)} b$. Let $\hat{a}$ be a representative of a. We can find a subset $J_0 \subseteq \kappa$ such that $|J_0| \leq |T| + \lambda$ and $\hat{a} \mathop{\smile\!\!\!\!\!\mid}_{(\hat{b}_i : i \in J_0)} (\hat{b}_i : i < \kappa)$ and hence $a \mathop{\smile\!\!\!\!\!\mid}_{(\hat{b}_i : i \in J_0)} b$. By symmetry and transitivity $a \mathop{\smile\!\!\!\!\!\mid}_{(b_i : i \in J_0)} b$. ⊣

COROLLARY 16.22. *Let T be simple. For any hyperimaginaries a, b, if a has length $\leq \lambda$, there is some hyperimaginary e of length $\leq |T| + \lambda$ such that $e \in \mathrm{dcl}^{\mathrm{heq}}(b)$ and $a \mathop{\smile\!\!\!\!\!\mid}_e b$.*

PROOF. By Lemma 15.4 there is a sequence $(b_i : i \in I)$ of countable hyperimaginaries b_i such that $b \sim (b_i : i \in I)$. By Proposition 16.21 there is some $J \subseteq I$ such that $|J| \leq |T| + \lambda$ and $a \mathop{\smile\!\!\!\!\!\mid}_{(b_i : i \in J)} (b_i : i \in I)$. Then $e = (b_i : i \in J)$ satisfies the requirements. ⊣

PROPOSITION 16.23. *Let T be simple, let e be a hyperimaginary, and let $\pi(x, y)$ be a set of formulas over $\emptyset$. Then $\pi(x, a)$ divides over e if and only if for any Morley sequence $(a_i : i < \omega)$ in $\mathrm{tp}(a/e)$, $\bigcup_{i<\omega} \pi(x, a_i)$ is inconsistent.*

PROOF. It is an adaptation of the proof of Proposition 5.15. We may assume $\pi(x, y) = \{\varphi(x, y)\}$. Let $(a_i : i < \omega)$ be a Morley sequence in $\mathrm{tp}(a/e)$ and assume $\varphi(x, a)$ divides over e and $\{\varphi(x, a_i) : i \in I\}$ is consistent. Let $\kappa \geq |T|$ be larger than the length of e and let $(I, <)$ be a linearly ordered set isomorphic to the reverse order of κ^+. There is a Morley sequence $a_I = (a_i : i < \kappa)$ in $\mathrm{tp}(a/e)$ such that $\{\varphi(x, a_i) : i \in I\}$ is consistent. Let c realize this set of formulas. By Proposition 16.21 there is some $J \subseteq I$ of cardinality $\leq \kappa$ such that $c \mathop{\smile\!\!\!\!\!\!\mid}_{ea_J} a_I$. We can find $i \in I$ such that $i < J$. Note that Lemma 5.14 is valid over a hyperimaginary and therefore $a_J \mathop{\smile\!\!\!\!\!\!\mid}_{e} a_i$. Proposition 4.9 is also valid over a hyperimaginary and hence $\varphi(x, a_i)$ divides over ea_J. But then $c \not\mathop{\smile\!\!\!\!\!\!\mid}_{ea_J} a_I$, a contradiction. ⊣

PROPOSITION 16.24 (Independence Theorem). *Let T be simple, and let a, b, c, d be hyperimaginaries such that $a \mathop{\smile\!\!\!\!\!\!\mid}_M b$, $c \mathop{\smile\!\!\!\!\!\!\mid}_M a$, $d \mathop{\smile\!\!\!\!\!\!\mid}_M b$, and $c \equiv_M d$. Then there is some hyperimaginary $e \mathop{\smile\!\!\!\!\!\!\mid}_M ab$ such that $e \equiv_{Ma} c$ and $e \equiv_{Mb} d$.*

PROOF. We may assume that c and d are real tuples (replace c by a representative $\hat{c}$ such that $\hat{c} \mathop{\smile\!\!\!\!\!\!\mid}_c Ma$ and then replace d by some representative $\hat{d}$ such that $\hat{c} \equiv_M \hat{d}$ and $\hat{d} \mathop{\smile\!\!\!\!\!\!\mid}_{Md} b$). Choose representatives $\hat{a}, \hat{b}$ of a and b such that $\hat{a} \mathop{\smile\!\!\!\!\!\!\mid}_M \hat{b}$. Consider $\mathrm{tp}(c/aM)$ and $\mathrm{tp}(d/bM)$ as partial types over $M\hat{a}$ and $M\hat{b}$ respectively. They can be extended to complete types $p(x) \in S(M\hat{a})$ and $q(x) \in S(M\hat{b})$ which do not fork over M. Note that $p \restriction M = q \restriction M$. By the Independence Theorem for ordinary types there is some $\hat{e} \models p \cup q$ such that $\hat{e} \mathop{\smile\!\!\!\!\!\!\mid}_M \hat{a}\hat{b}$. Then $\hat{e}$ is a representative of the hyperimaginary e we are seeking. ⊣

DEFINITION 16.25. SU-*rank* can be extended to hyperimaginaries in a natural way by the following conditions:

1. $\mathrm{SU}(a/e) \geq 0$.
2. $\mathrm{SU}(a/e) \geq \alpha + 1$ if and only if $\mathrm{SU}(a/eb) \geq \alpha$ for some hyperimaginary b such that $a \not\mathop{\smile\!\!\!\!\!\!\mid}_e b$.
3. $\mathrm{SU}(a/e) \geq \alpha$ if and only if $\mathrm{SU}(a/e) \geq \beta$ for all $\beta < \alpha$, if α is a limit ordinal.

One can check that $\{\alpha : \mathrm{SU}(a/e) \geq \alpha\}$ is an initial segment of the class of all ordinals numbers. We define $\mathrm{SU}(a/e) = \sup\{\alpha : \mathrm{SU}(a/e) \geq \alpha\}$. Hence, $\mathrm{SU}(a/e) = \infty$ if and only if $\mathrm{SU}(a/e) \geq \alpha$ for all ordinals α.

REMARK 16.26. *Let T be simple. For all hyperimaginaries a, b, e:*

1. $\mathrm{SU}(a/e) \geq \mathrm{SU}(a/eb)$.
2. *If $a \mathop{\smile\!\!\!\!\!\!\mid}_e b$, then* $\mathrm{SU}(a/e) = \mathrm{SU}(a/eb)$.

3. *If* $\mathrm{SU}(a/e) = \mathrm{SU}(a/eb) < \infty$, *then* $a \mathop{\smile\!\!\!\!\!\!|}_e b$.
4. *If* $\mathrm{SU}(ab/e) < \infty$, *then* $\mathrm{SU}(a/eb) + \mathrm{SU}(b/e) \leq \mathrm{SU}(ab/e) \leq \mathrm{SU}(a/eb) \oplus \mathrm{SU}(b/e)$.
5. *If* $b \in \mathrm{bdd}(ae)$, *then* $\mathrm{SU}(ab/e) = \mathrm{SU}(a/e)$.

PROOF. It is just a straightforward adaptation of the proofs of propositions 13.11, 13.16, and 13.18. ⊣

PROPOSITION 16.27. *Let T be simple. For all hyperimaginaries a, e:*

1. *If T is supersimple and a is finitary, then* $\mathrm{SU}(a/e) < \infty$.
2. $\mathrm{SU}(a/e) = 0$ *if and only if* $a \in \mathrm{bdd}(e)$.
3. $\mathrm{SU}(a/e) = \infty$ *if and only if there is a sequence* $(a_i : i < \omega)$ *of hyperimaginaries a_i such that* $a \not\mathop{\smile\!\!\!\!\!\!|}_{ea_{<i}} a_i$ *for all* $i < \omega$.
4. *If* $a = (a_i : i \in I)$ *and* $\mathrm{SU}(a/e) < \infty$, *then* $a \in \mathrm{bdd}(e, (a_i : i \in I_0))$ *for some finite* $I_0 \subseteq I$.

PROOF. 1. Observe first that for any finite real tuple a, for any hyperimaginary e we may choose a representative $\hat{e}$ of e such that $\hat{e} \mathop{\smile\!\!\!\!\!\!|}_e a$ and then, by Remark 16.26, $\mathrm{SU}(a/e) = \mathrm{SU}(a/e\hat{e}) \leq \mathrm{SU}(a/\hat{e}) < \infty$. Now, if a is a finitary hyperimaginary, and we choose a representative $\hat{a}$ of a, a finite tuple of real elements, then $a \in \mathrm{dcl}^{\mathrm{heq}}(\hat{a})$ and, again by Remark 16.26, $\mathrm{SU}(a/e) \leq \mathrm{SU}(\hat{a}/e)$ and therefore $\mathrm{SU}(a/e) < \infty$.

2. $\mathrm{SU}(a/e) = 0$ means that $a \mathop{\smile\!\!\!\!\!\!|}_e b$ for all b, and this is clearly equivalent to $a \in \mathrm{bdd}(e)$.

3. The proof of Proposition 13.11 can be adapted.

4. If $a \not\mathop{\smile\!\!\!\!\!\!|}_{e(a_i:i\in I_0)} a$ for all finite $I_0 \subseteq I$, then using Proposition 16.20 we easily obtain a sequence $(b_i : i < \omega)$ of hyperimaginaries b_i such that $a \not\mathop{\smile\!\!\!\!\!\!|}_{b_{<i}} b_i$ for all $i < \omega$. ⊣

DEFINITION 16.28. For any hyperimaginary e, the group of *strong automorphisms over* e is the group $\mathrm{Autf}(\mathfrak{C}/e)$ generated by all $\mathrm{Aut}(\mathfrak{C}/M)$ where $e \in \mathrm{dcl}^{\mathrm{heq}}(M)$. The *Lascar strong type* over e of a hyperimaginary a is the orbit $\mathrm{Lstp}(a/e)$ of a under $\mathrm{Autf}(\mathfrak{C}/e)$. Hence $\mathrm{Lstp}(a/e) = \mathrm{Lstp}(b/e)$ if and only if for some $n < \omega$ there are models M_i and hyperimaginaries a_i such that $e \in \mathrm{dcl}^{\mathrm{heq}}(M_i)$ and

$$a = a_0 \equiv_{M_0} a_1 \equiv_{M_1} a_2 \equiv_{M_3} \cdots \equiv_{M_n} a_{n+1} = b.$$

The equality of Lascar strong type for hyperimaginaries is defined as expected:

$$a \overset{\mathrm{Ls}}{\equiv}_e b \text{ if and only if } \mathrm{Lstp}(a/e) = \mathrm{Lstp}(b/e).$$

REMARK 16.29. *For any hyperimaginaries a, b, e:*

1. $a \overset{\mathrm{Ls}}{\equiv}_e b$ *if and only if* $\hat{a} \overset{\mathrm{Ls}}{\equiv}_e \hat{b}$ *for some representatives* $\hat{a}, \hat{b}$ *of* a, b.
2. *If* $a \equiv_M b$ *and* $e \in \mathrm{dcl}^{\mathrm{heq}}(M)$ *then for some hyperimaginary c, there are infinite e-indiscernible sequences I, J, such that $a, c \in I$ and $c, b \in J$.*

3. *If $a, b \in I$ for some infinite e-indiscernible sequence I, then $a \equiv_M b$ for some model M such that $e \in \mathrm{dcl}^{\mathrm{heq}}(M)$.*
4. *$a \overset{\mathrm{Ls}}{\equiv}_e b$ if and only if for some $n < \omega$ there are infinite e-indiscernible sequences I_i and hyperimaginaries a_i such that $a = a_0$, $a_i, a_{i+1} \in I_i$ and $a_{n+1} = b$.*

PROOF. 1 is clear. For 2, take representatives $\hat{a}, \hat{b}$ of a, b such that $\hat{a} \equiv_M \hat{b}$. By point 2 of Lemma 9.12, there is some $\hat{c}$ and some M-indiscernible (hence e-indiscernible) sequences I, J such that $\hat{a}, \hat{c} \in I$ and $\hat{c}, \hat{b} \in J$. The corresponding hyperimaginaries and sequences of hyperimaginaries are as required.

3. Let I be an e-indiscernible sequence such that $a, b \in I$. By Proposition 16.3, there are representatives $\hat{a}, \hat{b}, \hat{e}$ of a, b, e such that for some $\hat{e}$-indiscernible sequence J, $\hat{a}, \hat{b} \in J$. By point 1 of Lemma 9.12, $\hat{a} \equiv_M \hat{b}$ for some model $M \ni \hat{e}$. Then $e \in \mathrm{dcl}^{\mathrm{heq}}(M)$ and $a \equiv_M b$.

4 is a consequence of points 2 and 3. ⊣

DEFINITION 16.30. Le e be a hyperimaginary. A relation E is *type-definable over e* if it is type-definable and e-invariant. Note that this is equivalent to saying that E is type-definable over any representative of e. An equivalence class of a tuple in a type-definable over e equivalence relation is an *e-hyperimaginary*.

LEMMA 16.31. *Let e be a hyperimaginary. For any e-hyperimaginary h there is a hyperimaginary h' such that $h' \sim he$, that is, $\mathrm{Aut}(\mathfrak{C}/h') = \mathrm{Aut}(\mathfrak{C}/he)$. Moreover h' is e-bounded if h is e-bounded.*

PROOF. It is a generalization of Lemma 15.20, with a similar proof. Let $h = b_E$ where E is a type-definable over e equivalence relation. Let $\hat{e}$ be a representative of e, say $\hat{e}_G = e$, let $p(x) = \mathrm{tp}(\hat{e})$, and let $E = E(x, y; \hat{e})$. We define

$$F(xz, yu) \Leftrightarrow (G(z, u) \wedge p(z) \wedge p(u) \wedge E(x, y; z)) \vee xz = yu.$$

It is a 0-type-definable equivalence relation and it is easy to see that $h' = b\hat{e}_F$ is as required. ⊣

LEMMA 16.32. *Let T be simple. Let a, b, e be hyperimaginaries such that $a \mathop{\smile\!\!\!\!\!\!\mid}_e b$ and $a \equiv_e b$. There are representatives $\hat{a}, \hat{b}$ of a, b such that $\hat{a} \mathop{\smile\!\!\!\!\!\!\mid}_e \hat{b}$ and $\hat{a} \equiv_e \hat{b}$.*

PROOF. Start with a representative $\hat{a}$ of a such that $\hat{a} \mathop{\smile\!\!\!\!\!\!\mid}_a eb$. It follows that $\hat{a} \mathop{\smile\!\!\!\!\!\!\mid}_e b$. Next take some representative $\hat{b}$ of b such that $\hat{b} \equiv_e \hat{a}$ and $\hat{b} \mathop{\smile\!\!\!\!\!\!\mid}_{eb} \hat{a}$. Then $\hat{a} \mathop{\smile\!\!\!\!\!\!\mid}_e \hat{b}$. ⊣

LEMMA 16.33. *Let T be simple. Let a, b, e be hyperimaginaries such that $a \equiv_M b$ for some model M such that $e \in \operatorname{dcl}^{\text{heq}}(M)$. Then $a \equiv_N b$ for some model N such that $e \in \operatorname{dcl}^{\text{heq}}(N)$ and $ab \mathop{\smile\!\!\!\!\mid}_e N$.*

PROOF. We may assume that a, b are real tuples. Choose a model $M' \equiv_e M$ such that $M' \mathop{\smile\!\!\!\!\mid}_e M$ and then choose a model N such that $N \equiv_M M'$ and $\operatorname{tp}(N/Mab)$ is a coheir of $\operatorname{tp}(N/M)$. Then $N \mathop{\smile\!\!\!\!\mid}_e M$ and by Proposition 7.6 $N \mathop{\smile\!\!\!\!\mid}_M ab$. By transitivity, $N \mathop{\smile\!\!\!\!\mid}_e ab$. Since $\operatorname{tp}(N/Mab)$ does not split over M and $a \equiv_M b$, it follows that $a \equiv_N b$. ⊣

PROPOSITION 16.34. *Let T be simple. For any hyperimaginaries a, b, e such that $a \overset{\text{Ls}}{\equiv}_e b$ and $a \mathop{\smile\!\!\!\!\mid}_e b$ there is some model M such that $a \equiv_M b$, $ab \mathop{\smile\!\!\!\!\mid}_e M$ and $e \in \operatorname{dcl}^{\text{heq}}(M)$. Moreover $a, b \in I$ for some infinite M-indiscernible sequence I.*

PROOF. Fix $n < \omega$, fix models M_i for $i \leq n$ and sequences a_i for $i \leq n+1$ such that $e \in \operatorname{dcl}^{\text{heq}}(M_i)$ and

$$a = a_0 \equiv_{M_0} a_1 \equiv_{M_1} a_2 \equiv_{M_3} \cdots \equiv_{M_n} a_{n+1} = b.$$

By Lemma 16.33 we may assume that $a_i a_{i+1} \mathop{\smile\!\!\!\!\mid}_e M_i$. Hence we can also assume that $M_i \mathop{\smile\!\!\!\!\mid}_e a_0, \dots, a_{n+1}(M_j : j < i)$. It follows that

$$a_0, \dots, a_{n+1} \mathop{\smile\!\!\!\!\mid}_e M_0, \dots, M_n.$$

Note that $a_0 \mathop{\smile\!\!\!\!\mid}_{M_0,\dots,M_n} a_{n+1}$. Let $b_0 = a_0$, let $b_{n+1} = a_{n+1}$ and for $1 \leq i \leq n$ choose $b_i \equiv_{M_0,\dots,M_n} a_i$ and such that $b_i \mathop{\smile\!\!\!\!\mid}_{M_0,\dots,M_n} b_{n+1}(b_j : j < i)$ for any $i \leq n$. It follows that $(b_j : j \leq n+1)$ is independent over $M_0, \dots, M_n$. Note that

$$a = b_0 \equiv_{M_0} b_1 \equiv_{M_1} b_2 \equiv_{M_3} \cdots \equiv_{M_n} b_{n+1} = b.$$

Since $b_j \mathop{\smile\!\!\!\!\mid}_e M_0, \dots, M_n$, the sequence $(b_j : j \leq n+1)$ is also independent over e and moreover $b_0, \dots, b_{n+1} \mathop{\smile\!\!\!\!\mid}_e M_0, \dots, M_n$. Hence we can proceed by induction on n. The case $n = 0$ is clear and using the induction hypothesis it is enough now to consider the case $n = 1$.

We have $a = b_0 \equiv_{M_0} b_1 \equiv_{M_1} b_2 = b$. Let $f \in \operatorname{Aut}(\mathfrak{C}/M_0)$ such that $f(b_1) = b_0$. Then $M_1 \mathop{\smile\!\!\!\!\mid}_{M_0} b_1 b_2$, $f(M_1) \mathop{\smile\!\!\!\!\mid}_{M_0} b_0$ and $b_0 \mathop{\smile\!\!\!\!\mid}_{M_0} b_1 b_2$ and by the Independence Theorem (Proposition 16.24) there is a model N such that $N \mathop{\smile\!\!\!\!\mid}_{M_0} b_0 b_1 b_2$, $N \equiv_{M_0 b_1 b_2} M_1$ and $N \equiv_{M_0 b_0} f(M_1)$. Then $b_0 N \equiv b_0 f(M_1) \equiv b_1 M_1 \equiv b_2 M_1 \equiv b_2 N$ and clearly $b_0 b_2 \mathop{\smile\!\!\!\!\mid}_e N$.

The last assertion follows from Proposition 10.11 since $a \mathop{\smile\!\!\!\!\mid}_M b$ and by Lemma 16.32 we can assume a, b are real tuples. ⊣

PROPOSITION 16.35. *If T is simple, then for any hyperimaginaries $a = \hat{a}_F$, $b = \hat{b}_F$, and e, the following are equivalent:*

1. $a \overset{\text{Ls}}{\equiv}_e b$.

2. *For some hyperimaginary c there are infinite c-indiscernible sequences I, J, such that $a, c \in I$ and $c, b \in J$.*
3. $a \equiv_{\mathrm{bdd}(e)} b$.
4. *$E(\hat{a}, \hat{b})$ for every bounded e-type-definable equivalence relation $E \supseteq F$.*

PROOF. By Remark 16.29 it is clear that 2 implies 1 and that 1 implies 3. To prove 1 $\Rightarrow$ 2 find c such that $c \mathop{\smile\!\!\!\!\!\mid}_e ab$ and $a \stackrel{\mathrm{Ls}}{\equiv}_e c$ and then use Proposition 16.34. To find such c one needs to adapt the proof of Lemma 10.2, but this is straightforward.

3 $\Rightarrow$ 4. $\hat{a}_E$ is an e-hyperimaginary and by Lemma 16.31 there is a hyperimaginary c such that $c \sim e, \hat{a}_E$. Since $\hat{a}_E$ is e-bounded, $c \in \mathrm{bdd}(e)$. By 3, there is some $f \in \mathrm{Aut}(\mathfrak{C}/\mathrm{bdd}(e))$ such that $f(a) = b$. Then $F(f(\hat{a}), \hat{b})$ and therefore $E(f(\hat{a}), \hat{b})$. Since $f(c) = c$, $\hat{b}_E = f(\hat{a}_E) = \hat{a}_E$ and hence $E(\hat{a}, \hat{b})$.

4 $\Rightarrow$ 1. Let $E(x, y)$ be defined as $x_F \stackrel{\mathrm{Ls}}{\equiv}_e y_F$. The condition in 2 can be expressed by a partial type $\Phi(x, y, \hat{e})$ over any representative $\hat{e}$ of e. Hence $\Phi(x, y, \hat{e})$ defines E, which is a bounded e-type-definable equivalence relation $E \supseteq F$. By 4, $E(\hat{a}, \hat{b})$ and hence $a \stackrel{\mathrm{Ls}}{\equiv}_e b$. $\dashv$

COROLLARY 16.36. *Let T be simple. Let X be the set of all finite subsets of I and let $e = (e_J : J \in X)$ be a sequence of hyperimaginaries such that $e_J \in \mathrm{dcl}^{\mathrm{heq}}(e_{J'})$ whenever $J \subseteq J'$. For all hyperimaginaries a, b: $a \stackrel{\mathrm{Ls}}{\equiv}_e b$ if and only if $a \stackrel{\mathrm{Ls}}{\equiv}_{e_J} b$ for all $J \in X$.*

PROOF. By Proposition 16.35 and compactness. $\dashv$

COROLLARY 16.37 (Independence Theorem for hyperimaginary Lascar strong types). *Let T be simple, let a_1, a_2, c_1, c_2, e be hyperimaginaries such that $c_1 \mathop{\smile\!\!\!\!\!\mid}_e c_2$, $a_1 \mathop{\smile\!\!\!\!\!\mid}_e c_1$, $a_2 \mathop{\smile\!\!\!\!\!\mid}_e c_2$ and $a_1 \equiv_{\mathrm{bdd}(e)} a_2$. Then there is some hyperimaginary $a \mathop{\smile\!\!\!\!\!\mid}_e c_1c_2$ such that $a \equiv_{\mathrm{bdd}(e)c_1} a_1$ and $a \equiv_{\mathrm{bdd}(e)c_2} a_2$.*

PROOF. By Proposition 16.35, $a_1 \stackrel{\mathrm{Ls}}{\equiv}_e a_2$ and then by Proposition 16.34 $a_1 \equiv_M a_2$ for some model M such that $a_1a_2 \mathop{\smile\!\!\!\!\!\mid}_e M$ and $e \in \mathrm{dcl}^{\mathrm{heq}}(M)$. We can assume that $M \mathop{\smile\!\!\!\!\!\mid}_{a_1a_2e} c_1c_2$ and therefore $M \mathop{\smile\!\!\!\!\!\mid}_e c_1c_2a_1a_2$. It follows that $c_1 \mathop{\smile\!\!\!\!\!\mid}_M c_2$, $a_1 \mathop{\smile\!\!\!\!\!\mid}_M c_1$ and $a_2 \mathop{\smile\!\!\!\!\!\mid}_M c_2$. By Proposition 16.24 there is some $a \mathop{\smile\!\!\!\!\!\mid}_M c_1c_2$ such that $a \equiv_{Mc_1} a_1$ and $a \equiv_{Mc_2} a_2$. Clearly $a \mathop{\smile\!\!\!\!\!\mid}_e c_1c_2$, $a \equiv_{\mathrm{bdd}(e)c_1} a_1$ and $a \equiv_{\mathrm{bdd}(e)c_2} a_2$. $\dashv$

COROLLARY 16.38. *Let T be simple, let e be a hyperimaginary, and let $\varphi(x, y), \psi(x, z) \in L$. Assume $\varphi(x, a) \wedge \psi(x, b)$ does not fork over e. If $b \stackrel{\mathrm{Ls}}{\equiv}_e b'$, $a \mathop{\smile\!\!\!\!\!\mid}_e b$, and $a \mathop{\smile\!\!\!\!\!\mid}_e b'$, then $\varphi(x, a) \wedge \psi(x, b')$ does not fork over e.*

PROOF. Choose $c \mathop{\smile\!\!\!\!\!\mid}_e ab$ such that $c \models \varphi(x, a) \wedge \psi(x, b)$. Then $a \mathop{\smile\!\!\!\!\!\mid}_e c$, $b' \mathop{\smile\!\!\!\!\!\mid}_e a$, $b \mathop{\smile\!\!\!\!\!\mid}_e c$, and $b \stackrel{\mathrm{Ls}}{\equiv}_e b'$. We can apply the Independence Theorem 16.37 thus obtaining some $b'' \mathop{\smile\!\!\!\!\!\mid}_e ac$ such that $b'' \equiv_{ae} b'$ and $b'' \equiv_{ce} b$. We choose now c' such that $b''c \equiv_{ae} b'c'$. It is easy to see that $c' \mathop{\smile\!\!\!\!\!\mid}_e ab'$ and $c' \models \varphi(x, a) \wedge \psi(x, b')$. ⊣

COROLLARY 16.39. *Let T be simple, let $\Sigma(x)$ be a partial type representing a complete type over a hyperimaginary e, and let $\pi(x)$ be a partial type. Then $\Sigma(x) \cup \pi(x)$ does not fork over e if and only if $\mathrm{D}(\Sigma(x), \Delta, k) = \mathrm{D}(\Sigma(x) \cup \pi(x), \Delta, k)$ for all finite Δ, k.*

PROOF. It is an easy adaptation of the proof of Proposition 5.22 once we know that forking over hyperimaginaries has all the good properties. ⊣

LEMMA 16.40. *Let $\varphi(x, y) \in L$, and $n, k < \omega$. For any hyperimaginary e, for any ordinals α, β,*

$$\{(a, b) : a, b \text{ are tuples of length } \alpha, \beta \text{ and } \mathrm{D}(\mathrm{tp}(a/eb), \varphi, k) \geq n\}$$

is type-definable over any representative of e.

PROOF. Let $\hat{e}$ be a representative of e, say $e = \hat{e}_E$. As in the proof of Remark 3.14, $\mathrm{D}(\mathrm{tp}(a/eb), \varphi, k) \geq n$ if and only there is a tree $(a_s : s \in \omega^{\leq n})$ such that for each $f \in \omega^n$, $\mathrm{tp}(a/eb)(x) \cup \{\varphi(x, a_{f\restriction i+1}) : i < n\}$ is consistent and for each $i < n$, for each $s \in \omega^i$, $\{\varphi(x, a_{s^\frown j}) : j < \omega\}$ is k-inconsistent. Now consistency of $\mathrm{tp}(a/eb)(x) \cup \{\varphi(x, a_{f\restriction i+1}) : i < n\}$ can be expressed by:

$$\exists x (\bigwedge_{i<n} \varphi(x, a_{f\restriction i+1}) \wedge \exists z (E(z, \hat{e}) \wedge \bigwedge_{\psi(x,y,z) \in L} (\psi(x, b, \hat{e}) \leftrightarrow \psi(a, b, z)).$$

⊣

COROLLARY 16.41. *Let T be simple, let e be a hyperimaginary, and fix a type $p(x) \in S(e)$. For every ordinal α,*

$$\{(a, b) : a \models p,\ b \text{ is of length } \alpha \text{ and } a \mathop{\smile\!\!\!\!\!\mid}_e b\}$$

is type-definable over any representative of e.

PROOF. For each $\varphi(x, y) \in L$, for each $k < \omega$, let $n_{\varphi k} = \mathrm{D}(p(x), \varphi, k)$. By Corollary 16.39, for any $a \models p$, $a \mathop{\smile\!\!\!\!\!\mid}_e b$ if and only if $\mathrm{D}(\mathrm{tp}(a/eb), \varphi, k) \geq n_{\varphi k}$ for all φ, k. The result follows then by Lemma 16.40. ⊣

COROLLARY 16.42. *Let T be simple, let e be a hyperimaginary, and fix $p(x) \in S(e)$.*

1. *For any $n < \omega$*

   $$\{(a_1, \ldots, a_n) : (a_1, \ldots, a_n) \text{ is } e\text{-independent and } a_i \models p \text{ for } i = 1, \ldots, n\}$$

   is type-definable over any representative of e.
2. *For any linearly ordered set I,*

   $$\{(a_i : i \in I) : (a_i : i \in I) \text{ is a Morley sequence in } p\}$$

   is type-definable over any representative of e.

PROOF. Like the proof of Corollary 5.24, using Lemma 16.40 and Corollary 16.39. ⊣

Chapter 17

CANONICAL BASES REVISITED

DEFINITION 17.1. Let $p(x) \in S_E(e)$ be a complete hyperimaginary type over the hyperimaginary e, that is, $p(x) = \text{tp}(a/e)$ where a, e are hyperimaginaries and E is the sort of a. We say that $p(x)$ is an *amalgamation base* if the Independence Theorem is true for $p(x)$: for any hyperimaginaries c_1, c_2 such that $c_1 \mathop{\smile\!\!\!\!\!\mid}_e c_2$, for any hyperimaginaries a_1, a_2 such that $\text{tp}(a_1/e) = p(x) = \text{tp}(a_2/e)$, $a_1 \mathop{\smile\!\!\!\!\!\mid}_e c_1$, and $a_2 \mathop{\smile\!\!\!\!\!\mid}_e c_2$, there is some hyperimaginary $a \mathop{\smile\!\!\!\!\!\mid}_e c_1c_2$ such that $a \equiv_{ec_1} a_1$ and $a \equiv_{ec_2} a_2$. It is easy to see that (if T is simple) we can always check this condition assuming c_1, c_2 are real tuples enumerating models M_1, M_2 such that $e \subset \text{dcl}^{\text{heq}}(M_i)$ for $i = 1, 2$.

As shown in Proposition 16.24 and in Corollary 16.37, in a simple theory any type over a model and any type over a hyperimaginary of the form $\text{bdd}(e)$ is an amalgamation base. Hence Lascar strong types (more precisely, their corresponding sets of formulas) in simple theories are amalgamation bases.

Let $p(x) \in S_E(e)$, and let $d \in \text{dcl}^{\text{heq}}(e)$. By $p(x) \restriction d$ we refer to the type $\text{tp}(b/d)$ where b is an arbitrary realization of p. It is well defined. Note that if $q(x)$ is another complete type of sort E and d is also definable over its domain, then the consistency of $p(x) \cup q(x)$ implies $p \restriction d = q \restriction d$. In particular, if $\mathfrak{p} \in S_E(\mathfrak{C})$ is a global type in the hyperimaginary sort E, then $\mathfrak{p} \restriction e$ is well defined for any hyperimaginary e.

The notion of *stationary* type also makes sense for hyperimaginary types. $p(x) \in S_E(e)$ is stationary if there is a unique global type $\mathfrak{p}(x)$ extending p (i.e., such that $\mathfrak{p} \restriction e = p$) that does not fork over e.

REMARK 17.2. *Let T be simple. For any hyperimaginaries a, e, $\text{tp}(a/e)$ is an amalgamation base if and only if $\text{tp}(a/e) \vdash \text{tp}(a/\text{bdd}(e))$.*

PROOF. Assume $\text{tp}(a/e)$ is an amalgamation base and $a' \equiv_e a$. We will show that $a' \equiv_{\text{bdd}(e)} a$. Notice that $a \mathop{\smile\!\!\!\!\!\mid}_e \text{bdd}(e)$, $a' \mathop{\smile\!\!\!\!\!\mid}_e \text{bdd}(e)$, and $\text{bdd}(e) \mathop{\smile\!\!\!\!\!\mid}_e \text{bdd}(e)$. We can amalgamate these types and obtain some a'' such that $a'' \mathop{\smile\!\!\!\!\!\mid}_e \text{bdd}(e)$, $a'' \equiv_{\text{bdd}(e)} a'$, and $a'' \equiv_{\text{bdd}(e)} a$. It follows that $a \equiv_{\text{bdd}(e)} a'$. The other direction follows from Corollary 16.37. ⊣

PROPOSITION 17.3. 1. *In a simple theory, any stationary type is an amalgamation base.*

2. *Let T be simple. Assume $p(x) \in S_E(e)$ is stationary and $q(x) \in S_E(d)$ is an amalgamation base. If p and q have a common nonforking extension, then q is stationary.*
3. *In a stable theory, all amalgamation bases are stationary.*

PROOF. 1. Given a_1, a_2, c_1, c_2, e and a stationary type $p(x)$ over e as in the definition of amalgamation base, any realization a of the unique nonforking extension of p over e, c_1, c_2 satisfies the requirements.

2. Assume there is a set A such that $d \in \mathrm{dcl}^{\mathrm{heq}}(A)$ and q has two different nonforking extensions $q_1(x), q_2(x)$ over A. Without loss of generality, $e \in \mathrm{dcl}^{\mathrm{heq}}(A)$ and $p = q_1$. Choose $A' \equiv_e A$ such that $A' \mathop{\smile\!\!\!\!\!\!\mid}_e A$ and let p' be the corresponding copy of p over A by an automorphism over e. Then p' is stationary and a nonforking extension of q. Since q is an amalgamation base, q_2 and p' have a common global nonforking extension over AA'. But p and p' also have a common nonforking extension over AA'. By stationarity of p' they coincide, and therefore $p = q_2$.

3. Let T be stable. By Lemma 16.17 we only need to consider the case of a type $p(x) \in S(e)$ of real variables. If e is a sequence of real parameters the result follows easily from Remark 17.2, since in T strong types are stationary. But, in any case, the general case follows from 2 since in T stable all amalgamation bases have nonforking stationary extensions, for instance nonforking extensions over a model. ⊣

REMARK 17.4. *Let T be simple. Assume a, b, e are hyperimaginaries and $b \in \mathrm{dcl}^{\mathrm{heq}}(ae)$. If $\mathrm{tp}(a/e)$ is an amalgamation base, then $\mathrm{tp}(b/e)$ is also an amalgamation base.*

PROOF. By Remark 17.2. ⊣

DEFINITION 17.5. Let $p(x) \in S_E(e)$ for a hyperimaginary e and let A be a class of hyperimaginaries. We say that p is *finitely satisfiable in A* if for every formula $\varphi(x) \in p(x)$ (considering $p(x)$ as a partial type, which we assume is closed under finite conjunctions, over some representative of e) there is some hyperimaginary $b_E \in A$ such that $\models \varphi(\hat{b})$ for every representative $\hat{b}$ of b_E. This definition is independent of the chosen representative of e.

PROPOSITION 17.6. *Let T be simple and let a, d, e be hyperimaginaries.*

1. *If $\mathrm{tp}(a/e)$ is finitely satisfiable in $\mathrm{dcl}^{\mathrm{heq}}(d)$ and $d \in \mathrm{dcl}^{\mathrm{heq}}(e)$, then $a \mathop{\smile\!\!\!\!\!\!\mid}_d e$.*
2. *If $\mathrm{tp}(a/e)$ is finitely satisfiable in $\mathrm{dcl}^{\mathrm{heq}}(e)$, then $\mathrm{tp}(a/e)$ is an amalgamation base.*

PROOF. 1. If $\hat{e}$ is a representative of e and $\varphi(x, \hat{e}) \in \mathrm{tp}(a/e)$ then $\models \varphi(\hat{b}, \hat{e})$ for all representatives $\hat{b}$ of some $b \in \mathrm{dcl}^{\mathrm{heq}}(d)$ and hence $\varphi(x, \hat{e})$ does not divide over d.

2. By Remark 17.2 it suffices to show that $\mathrm{tp}(a/e) \vdash \mathrm{tp}(a/\mathrm{bdd}(e))$. We use Proposition 16.35. Let $a = a'_F$, $b = b'_F$, let $E \supseteq F$ be a type-definable over

e bounded equivalence relation, assume $a \equiv_e b$, and let us prove $E(a', b')$. Choose a representative e' of e, and choose some $f \in \mathrm{Aut}(\mathfrak{C}/e)$ such that $f(a) = b$. Then $e'' = f(e')$ is also a representative of e. Let $\Sigma(x, y, u)$ be a partial type over $\emptyset$ such that $\Sigma(x, y, e')$ defines E. Then $\Sigma(x, y, e'')$ also defines E. Consider some $\varphi(x, y, u) \in \Sigma(x, y, u)$. We will check that $\models \varphi(a', b', e')$. Choose now by compactness some symmetric (in x, y) formula $\psi(x, y, u) \in \Sigma(x, y, u)$ such that

$$\psi(x, y_1, e') \wedge \psi(y_1, y_2, e') \wedge \psi(y_2, y_3, e') \wedge \\ \psi(y_3, y_4, e'') \wedge \psi(y_4, z, e') \vdash \varphi(x, z, e').$$

Since E is bounded, $\psi(x, y, e')$ is thick and we can choose a maximal sequence $c_1, \ldots, c_n \in \mathrm{dcl}^{\mathrm{heq}}(e)$ such that $\models \neg\psi(c_i', c_j', e')$ for all representatives c_i', c_j' of c_i, c_j, for all $i \neq j$.

Assume that $\models \psi(a'', c_i', e')$ for some representative a'' of a, for some representative c_i' of c_i, for some $i \leq n$. Then $f(a'')$ is a representative of b, $f(c_i')$ is a representative of c_i, and $\models \psi(f(a''), f(c_i'), e'')$. Since $F(x, y) \vdash \psi(x, y, e')$, by choice of ψ we obtain that $\models \varphi(a', b', e')$.

Assume now that $\models \neg\psi(a'', c_i', e')$ for any representative a'' of a, for any representative c_i' of c_i, for any $i \leq n$. Consider $\mathrm{tp}(a/ec_1, \ldots, c_n)$ as a partial type over $e', c_1', \ldots, c_n'$ for some representatives $c_1', \ldots, c_n'$ of $c_1, \ldots, c_n$. Notice that it is finitely satisfiable in $\mathrm{dcl}^{\mathrm{heq}}(e)$. By compactness there is some symmetric formula $\varepsilon(x, y) \in F(x, y)$ such that

$$\models \forall x' x_1 \ldots x_n (\varepsilon(x', a') \wedge \bigwedge_{i=1}^{n} \varepsilon(x_i, c_i') \rightarrow \bigwedge_{i=1}^{n} \neg\psi(x', x_i, e')).$$

If $\theta(y, z, z_1, \ldots, z_n)$ is the formula $\forall x' x_1 \ldots x_n (\varepsilon(x', y) \wedge \bigwedge_{i=1}^{n} \varepsilon(x_i, z_i) \rightarrow \bigwedge_{i=1}^{n} \neg\psi(x', x_i, z))$, then, by the way $\mathrm{tp}(a/ec_1, \ldots, c_n)$ has been constructed, $\exists y(\varepsilon(x, y) \wedge \theta(y, e', c_1', \ldots, c_n')) \in \mathrm{tp}(a/ec_1, \ldots, c_n)$ and therefore there is some $d \in \mathrm{dcl}^{\mathrm{heq}}(e)$ of sort F such that $\models \exists y(\varepsilon(d', y) \wedge \theta(y, e', c_1', \ldots, c_n'))$ for every representative d' of d. This contradicts the maximality of the sequence $c_1, \ldots, c_n$. ⊣

COROLLARY 17.7. *Let T be simple. If $(a_i : i \leq \omega)$ is an indiscernible sequence of hyperimaginaries, then $\mathrm{tp}(a_\omega/(a_i : i < \omega))$ is an amalgamation base.*

PROOF. It is a consequence of Proposition 17.6 since $\mathrm{tp}(a_\omega/(a_i : i < \omega))$ is finitely satisfiable in $\{a_i : i < \omega\}$. ⊣

PROPOSITION 17.8. *Let T be simple, let a_E be a hyperimaginary and let $D = \mathrm{dcl}^{\mathrm{heq}}(a) \cap \mathrm{bdd}(a_E)$. Then $\mathrm{tp}(a/D)$ is an amalgamation base. Moreover there is some 0-type-definable equivalence relation E' such that $a_{E'}$ is equivalent to some enumeration of D and hence $a_{E'} \in \mathrm{bdd}(a_E)$, $a_E \in \mathrm{dcl}^{\mathrm{heq}}(a_{E'})$, and $\mathrm{tp}(a/a_{E'})$ is an amalgamation base.*

PROOF. Let e be a hyperimaginary equivalent to some enumeration of D. Then $e \in \mathrm{dcl}^{\mathrm{heq}}(a)$ and therefore, by Proposition 15.6, $e \sim a_{E'}$ for some 0-type-definable equivalence relation E'. By Remark 17.2 it is enough to show that $\mathrm{tp}(a/e) \vdash \mathrm{tp}(a/\mathrm{bdd}(e))$. Let $a \equiv_e a'$. We will check that $a \overset{\mathrm{Ls}}{\equiv}_e a'$. By Proposition 16.35, it suffices to check that $F(a, a')$ for every bounded e-type-definable equivalence relation F. Let F be such an equivalence relation. By Lemma 16.31 there is an e-bounded hyperimaginary h such that $h \sim a_F e$. Since $h \in \mathrm{bdd}(e)$ and $e \in \mathrm{bdd}(a_E)$, by Remark 15.19 $h \in \mathrm{bdd}(a_E)$. Moreover $h \in \mathrm{dcl}^{\mathrm{heq}}(a_F, e)$ and $a_F \in \mathrm{dcl}^{\mathrm{heq}}(a, e) \subseteq \mathrm{dcl}^{\mathrm{heq}}(a)$, and hence $h \in \mathrm{dcl}^{\mathrm{heq}}(a)$. Therefore $h \in D \subseteq \mathrm{dcl}^{\mathrm{heq}}(e)$. If $f \in \mathrm{Aut}(\mathfrak{C}/e)$ is such that $f(a) = a'$, then $f(h) = h$ and therefore $a_F = f(a_F) = f(a)_F$, that is, $F(a, a')$. ⊣

DEFINITION 17.9. Let e be a hyperimaginary. A *canonical base of an amalgamation base* $p(x) \in S_E(e)$ is a smallest hyperimaginary $d \in \mathrm{dcl}^{\mathrm{heq}}(e)$ such that p does not fork over d and $p \restriction d$ is an amalgamation base. This means that if $d' \in \mathrm{dcl}^{\mathrm{heq}}(e)$ is another hyperimaginary such that p does not fork over d' and $p \restriction d'$ is an amalgamation base then $d \in \mathrm{dcl}^{\mathrm{heq}}(d')$. We will see that in simple theories all amalgamation bases have canonical bases.

REMARK 17.10. *Let T be simple. Assume $p(x) \in S(A)$ is stationary. By Lemma* 11.2, *its global nonforking extension $\mathfrak{p}$ is definable. Then p is an amalgamation base and two notions of canonical base may be applied to p. It follows from Proposition* 17.3 *that they are interdefinable.*

DEFINITION 17.11. Let $p(x)$ be an amalgamation base of hyperimaginary sort. The *amalgamation class of p* is the class $\mathcal{P}_p$ consisting of all global types $\mathfrak{p}(x)$ such that for some $n < \omega$ there are global types $(\mathfrak{p}_i(x) : i \leq n)$ such that $\mathfrak{p}_0$ is a nonforking extension of p, $\mathfrak{p} = \mathfrak{p}_n$ and for every $i < n$, $\mathfrak{p}_i$ and $\mathfrak{p}_{i+1}$ are nonforking extensions of a common amalgamation base $p_i(x) \subseteq \mathfrak{p}_i, \mathfrak{p}_{i+1}$. Note that $\mathcal{P}_p = \mathcal{P}_q$ if and only if $\mathcal{P}_p \cap \mathcal{P}_q \neq \emptyset$.

REMARK 17.12. *In a simple theory, if $p(x) \in S_E(e)$ is an amalgamation base over a hyperimaginary e, there is always an amalgamation base $q(x) \in S_E(M)$ where M is a model and $\mathcal{P}_q = \mathcal{P}_p$.*

PROOF. If $p(x) \in S_E(e)$ where e is a hyperimaginary, choose a model M such that $e \in \mathrm{dcl}^{\mathrm{heq}}(M)$ and set $q(x) = \mathfrak{p} \restriction M$ where $\mathfrak{p}$ is a global nonforking extension of p. ⊣

REMARK 17.13. *Let T be simple and let e be a hyperimaginary. If $p(x) \in S_E(e)$ is stationary, then the amalgamation class $\mathcal{P}_p$ is a singleton $\{\mathfrak{p}\}$ where $\mathfrak{p}$ is the unique global nonforking extension of p. Hence in stable theories amalgamation classes are singletons.*

PROOF. Let $p(x)$ be stationary and let $\mathfrak{p}$ be its global nonforking extension. If $q(x)$ is an amalgamation base and p, q have a common global nonforking extension, this extension is $\mathfrak{p}$ and by Proposition 17.3 $q(x)$ is stationary. Hence it is parallel to p and $\mathfrak{p}$ is its only global nonforking extension. ⊣

LEMMA 17.14. *Let T be simple, let a, b be hyperimaginaries and let $p(x) \in S_E(a)$ be an amalgamation base. If $q(x) \in S_E(b)$ is another amalgamation base and $\mathcal{P}_p = \mathcal{P}_q$, then for some $n < \omega$ there are automorphisms $f_0, \ldots, f_n \in \mathrm{Aut}(\mathfrak{C})$ such that if $p_i = p^{f_i}$, then $p_0 = p$, $p_i(x)$ and $p_{i+1}(x)$ have a common nonforking extension for all $i < n$, and also $p_n(x)$ and $q(x)$ have a common nonforking extension.*

PROOF. It is enough to prove that for all amalgamation bases $q(x) \in S_E(c)$, $r(x) \in S_E(d)$, if p and q have a common nonforking extension and q and r also have a common nonforking extension, then for some automorphism $f \in \mathrm{Aut}(\mathfrak{C})$, p and p^f have a common nonforking extension and p^f and r also have a common nonforking extension.

To check this, let us choose $b \equiv_c a$ such that $b \mathop{\smile\!\!\!\!\!\!|}_c ad$ and let $f \in \mathrm{Aut}(\mathfrak{C}/c)$ be such that $f(a) = b$. Since $b \equiv_c a$, p^f and q have a common nonforking extension $s_1(x) \in S_E(bc)$. Now let $s_2(x) \in S_E(cd)$ be a common nonforking extension of q and r. Since $b \mathop{\smile\!\!\!\!\!\!|}_c d$ and q is an amalgamation base, s_1 and s_2 have a common extension $s(x) \in S_E(bcd)$ which does not fork over c. Then s is a common nonforking extension of r and p^f. We finish the proof showing that also p and p^f have a common nonforking extension. On the one hand $s_1 \in S_E(bc)$ is a common nonforking extension of p^f and q. On the other hand q and p have a common nonforking extension $s_3(x) \in S_E(ac)$. Since $a \mathop{\smile\!\!\!\!\!\!|}_c b$, we see that s_1 and s_3 have a common extension $s'(x) \in S_E(abc)$ which does not fork over c. Clearly it is a common nonforking extension of p and p^f. ⊣

LEMMA 17.15. *Let T be simple, let R be a 0-type-definable symmetric relation on tuples of a given length. Assume that R is generically transitive, that is, whenever $R(a, b)$, $R(a, c)$, and $b \mathop{\smile\!\!\!\!\!\!|}_a c$, then $R(b, c)$. Let $R^n = R \circ R \circ \cdots \circ R$ (n-times) be the n-step composition of R. Then:*

1. *If $R(a, b)$, then for any tuple d there is some $c \equiv b$ such that $R(a, c)$, $R(b, c)$, $bd \mathop{\smile\!\!\!\!\!\!|}_a c$, and $ad \mathop{\smile\!\!\!\!\!\!|}_b c$.*
2. *$R^n \subseteq R^2$ for all $n \geq 2$.*
3. *If $R^2(a, b)$, then there is some $c \equiv b$ such that $R(a, c)$, $R(b, c)$, $b \mathop{\smile\!\!\!\!\!\!|}_a c$, and $a \mathop{\smile\!\!\!\!\!\!|}_b c$.*

In particular, since the transitive closure of R is R^2 it is 0-type-definable.

PROOF. 1. Let Σ be the set of all formulas $\varphi(x, y) \in L$ with a fixed tuple x of variables and let us define for tuples c of the length of x,

$$\bar{\mathrm{D}}(c/a) = (\mathrm{D}(\mathrm{tp}(c/a), \varphi, k) : \varphi \in \Sigma, k \in \omega).$$

We consider $\bar{\mathrm{D}}(c/a)$ as element of the partially ordered set $\omega^{\Sigma\times\omega}$ of all sequences $(n_{\varphi,k} : \varphi \in \Sigma, k \in \omega)$. Assume $R(a,b)$ and choose, with Corollary 3.15, some c with maximal $\bar{\mathrm{D}}(c/a)$ among all tuples $c \equiv b$ such that $R(a,c)$. Without loss of generality, $c \mathop{\smile\!\!\!\!\!\!\mid}_a bd$. Since R is generically transitive, $R(b,c)$. It is enough to prove that $c \mathop{\smile\!\!\!\!\!\!\mid}_b a$ since from this it follows $c \mathop{\smile\!\!\!\!\!\!\mid}_b ad$ by ordinary computations in the forking calculus. Assume $c \not\mathop{\smile\!\!\!\!\!\!\mid}_b a$. Then $\bar{\mathrm{D}}(c/b) > \bar{\mathrm{D}}(c/ab)$. Since $c \mathop{\smile\!\!\!\!\!\!\mid}_a b$, $\bar{\mathrm{D}}(c/ab) = \bar{\mathrm{D}}(c/a)$. Hence, $\bar{\mathrm{D}}(c/b) > \bar{\mathrm{D}}(c/a)$. Choose $c' \equiv_b c$ such that $c' \mathop{\smile\!\!\!\!\!\!\mid}_b a$. Then $R(c',b)$ and, by generic transitivity, $R(c',a)$. Moreover $c' \equiv b$. Using the fact that $c' \mathop{\smile\!\!\!\!\!\!\mid}_b a$, we see that $\bar{\mathrm{D}}(c'/a) \geq \bar{\mathrm{D}}(c'/ab) = \bar{\mathrm{D}}(c'/b) = \bar{\mathrm{D}}(c/b) > \bar{\mathrm{D}}(c/a)$, contradicting the maximality of $\bar{\mathrm{D}}(c/a)$.

2. It is enough to prove that $R^3(a,b)$ implies $R^2(a,b)$. Assume a, a', b', b is an R-path connecting a with b in three steps. Apply 1 to a', b' and the additional tuple $d = ab$. We obtain some $c \equiv b'$ such that $R(a',c)$, $R(b',c)$, $b'ab \mathop{\smile\!\!\!\!\!\!\mid}_{a'} c$ and $a'ab \mathop{\smile\!\!\!\!\!\!\mid}_{b'} c$. Since R is generically transitive, $R(a,c)$ and $R(b,c)$.

3. Assume $R(a,c)$ and $R(b,c)$. By 1 applied to a, c with $d = b$ we find $c' \equiv c$ such that $c' \mathop{\smile\!\!\!\!\!\!\mid}_a cb$ and $c' \mathop{\smile\!\!\!\!\!\!\mid}_c ab$. Since R is generically transitive, $R(b,c')$. Now apply 1 to b, c' with $d = a$ obtaining $c'' \equiv b$ such that $c'' \mathop{\smile\!\!\!\!\!\!\mid}_b c'a$ and $c'' \mathop{\smile\!\!\!\!\!\!\mid}_{c'} ba$. By generic transitivity of R, $R(a,c'')$. It is easy to check that $c'' \mathop{\smile\!\!\!\!\!\!\mid}_a b$ and $c'' \mathop{\smile\!\!\!\!\!\!\mid}_b a$. ⊣

THEOREM 17.16. *In a simple theory, for every amalgamation base $p(x)$ of hyperimaginary sort there is a hyperimaginary e such that for every automorphism $f \in \mathrm{Aut}(\mathfrak{C})$, $f(e) = e$ if and only if f fixes setwise the amalgamation class $\mathcal{P}_p$.*

PROOF. By Remark 17.12, it is enough to consider an amalgamation base of the form $p(x) \in S_E(a)$ where a is a real tuple. Write $p(x) = p(x,a)$ with $p(x,y)$ over $\emptyset$ and consider the binary relation R on realizations of $\mathrm{tp}(a)$ defined by: $R(a_1,a_2)$ if and only if $p(x,a_1)$ and $p(x,a_2)$ have a common nonforking extension. It is symmetric. For each $\varphi(x,y) \in L$, for each $k < \omega$ let $n_{\varphi,k} = \mathrm{D}(p(x,a),\varphi,k)$. Then it is easy to see that R is type-definable by the partial type (over $\emptyset$) expressing that a_1, a_2 realize $\mathrm{tp}(a)$ and for all $\varphi \in L$, for all $k < \omega$, $\mathrm{D}(p(x,a_1) \cup p(x,a_2),\varphi,k) \geq n_{\varphi,k}$.

R satisfies all the remaining conditions of Lemma 17.15 and therefore its transitive closure F is also type-definable. Note that F is an equivalence relation on realizations of $\mathrm{tp}(a)$ and by Lemma 17.14 $F(a,b)$ holds if and only if $\mathcal{P}_{p(x,b)} = \mathcal{P}_{p(x,a)}$. By Lemma 15.5 we can extend F to a 0-type-definable equivalence relation on all sequences of the length of a. Hence we can consider the hyperimaginary $e = a_F$. It is clear that e satisfies the requirements. ⊣

LEMMA 17.17. *Let T be simple, let $p(x) \in S_E(a)$ be an amalgamation base and let e be a hyperimaginary equivalent to the amalgamation class $\mathcal{P}_p$ as in*

Theorem 17.16. *If $q(x) \in S_E(b)$ is an amalgamation base, $\mathcal{P}_p = \mathcal{P}_q$, and $a \mathop{\smile\!\!\!\!\!\!|}_e b$ then p and q have a common nonforking extension.*

PROOF. By enlarging a, b if necessary, we may assume they are real tuples of the same length. By Lemma 17.14 we may now find $n < \omega$, types $p_i(x, y)$ over $\emptyset$ and tuples a_i for $i \leq n$ such that $p(x) = p_0(x, a_0)$, $q(x) = p_n(x, a_n)$, and $p_i(x, a_i)$ and $p_{i+1}(x, a_{i+1})$ are amalgamation bases with a common nonforking extension for all $i < n$. We may apply 17.15 to the relation R defined by $R(b_1, b_2)$ if and only if there are $i, j \leq n$ such that $b_1 \equiv a_i$, $b_2 \equiv a_j$ and $p_i(x, b_1)$, $p_j(x, b_2)$ have a common nonforking extension. Hence there is an amalgamation base $r(x) \in S_E(c)$ such that $p(x), r(x)$ have a common nonforking extension, $r(x), q(x)$ have a common nonforking extension, $c \mathop{\smile\!\!\!\!\!\!|}_a b$, and $c \mathop{\smile\!\!\!\!\!\!|}_b a$. Since $a \mathop{\smile\!\!\!\!\!\!|}_e b$ and $e \in \mathrm{dcl}^{\mathrm{heq}}(b) \cap \mathrm{dcl}^{\mathrm{heq}}(c)$ from this it follows that $a \mathop{\smile\!\!\!\!\!\!|}_c b$. By amalgamating these types we conclude that $p(x)$ and $q(x)$ have a common nonforking extension. ⊣

THEOREM 17.18. *Let T be simple, let $p(x) \in S_E(a)$ be an amalgamation base, let $e \in \mathrm{dcl}^{\mathrm{heq}}(a)$ be a hyperimaginary equivalent to the amalgamation class $\mathcal{P}_p$ as in Theorem 17.16, and let $p_0 = p \restriction e$.*

1. *Any $\mathfrak{p}(x) \in \mathcal{P}_p$ is a nonforking extension of p_0.*
2. *p_0 is an amalgamation base and $\mathcal{P}_p = \mathcal{P}_{p_0}$.*
3. *If $q(x) \in S_E(b)$ is an amalgamation base and $p(x), q(x)$ have a common nonforking extension, then $e \in \mathrm{dcl}^{\mathrm{heq}}(b)$.*
4. *If $p(x)$ and $q(x) \in S_E(b)$ have a common nonforking extension, then $e \in \mathrm{bdd}(b)$.*
5. *If $b \in \mathrm{dcl}^{\mathrm{heq}}(a)$, then $p(x)$ does not fork over b if and only if $e \in \mathrm{bdd}(b)$.*

PROOF. 1. Choose $b \equiv_e a$ such that $b \mathop{\smile\!\!\!\!\!\!|}_e a$. Let $f \in \mathrm{Aut}(\mathfrak{C}/e)$ be such that $f(a) = b$ and let $q(x) = p^f$. Then $\mathcal{P}_q = \mathcal{P}_p$ and by Lemma 17.17 $p(x)$ and $q(x)$ have a common global nonforking extension. Hence there is a common realization c of p and q such that $c \mathop{\smile\!\!\!\!\!\!|}_a b$. Since $e \in \mathrm{dcl}^{\mathrm{heq}}(b)$ by symmetry and transitivity, $c \mathop{\smile\!\!\!\!\!\!|}_e a$ and therefore p does not fork over e. Hence any global nonforking extension of p does not fork over e.

If $\mathfrak{p} \in \mathcal{P}_p$ we can choose an amalgamation base $q(x)$ such that $\mathfrak{p}$ is a nonforking extension of $q(x)$. Since $\mathcal{P}_p = \mathcal{P}_q$, e is a canonical base of q and hence $\mathfrak{p}$ does not fork over e.

2. Let $\mathcal{P} = \mathcal{P}_{p_0}$. Choose a model M and $q(x) \in S_E(M)$ such that $e \in \mathrm{dcl}^{\mathrm{heq}}(M)$ and $\mathcal{P}_q = \mathcal{P}$. Then $p_0 = q(x) \restriction e$. Since $\mathrm{bdd}(e) \subseteq \mathrm{dcl}^{\mathrm{heq}}(M)$ and $p_0' = q(x) \restriction \mathrm{bdd}(e)$ is an amalgamation base, by 1 $\mathcal{P}_{p_0'} = \mathcal{P}$. We will see now that all extensions of p_0 over $\mathrm{bdd}(e)$ have the same property. Let $p_0''(x) = \mathrm{tp}(c/\mathrm{bdd}(e))$ be any such extension and choose b such that $p_0' = \mathrm{tp}(b/\mathrm{bdd}(e))$. Then $f(b) = c$ for some $f \in \mathrm{Aut}(\mathfrak{C}/e)$. Clearly, f fixes

setwise $\mathrm{bdd}(e)$ and $(p_0')^f = p_0''$. Since $f(e) = e$, f fixes setwise $\mathcal{P}$, and thus $\mathcal{P} = \mathcal{P}_{p_0''}$, as we wanted to show.

Next we check that p_0 is an amalgamation base. Let M, N be models such that $e \in \mathrm{dcl}^{\mathrm{heq}}(M) \cap \mathrm{dcl}^{\mathrm{heq}}(N)$ and $M \mathop{\smile\!\!\!\!\!\mid}_e N$. Assume $q_1(x) \in S_E(M)$, $q_2(x) \in S_E(N)$ are nonforking extensions of p_0. Note that $q_1(x)$ is a nonforking extension of $q_1(x) \restriction \mathrm{bdd}(e)$ and hence $\mathcal{P}_{q_1} = \mathcal{P}$. For similar reasons, $\mathcal{P}_{q_2} = \mathcal{P}$. By Lemma 17.17 $q_1(x), q_2(x)$ have a common nonforking extension which, by symmetry and transitivity, does not fork over e.

3 is clear since $\mathcal{P}_p = \mathcal{P}_q$, and 4 follows from 3 because $\mathcal{P}_p = \mathcal{P}_{q'}$ for some extension $q'(x)$ of $q(x)$ over $\mathrm{bdd}(b)$.

5. One direction follows from 4. For the opposite direction use 1 to see that p does not fork over e. ⊣

COROLLARY 17.19. *Let T be simple and let $p(x) \in S_E(a)$ be an amalgamation base with amalgamation class $\mathcal{P}_p$. The following are equivalent for any hyperimaginary e:*

1. *e is a canonical base of $p(x)$.*
2. *For any $f \in \mathrm{Aut}(\mathfrak{C})$, $f(e) = e$ if and only if $f(\mathcal{P}_p) = \mathcal{P}_p$.*

PROOF. $1 \Rightarrow 2$. Let $\mathfrak{p}$ be a global nonforking extension of p. If $f(e) = e$, then $\mathfrak{p}$ and $\mathfrak{p}^f$ are nonforking extensions of the amalgamation base $p \restriction e$ and therefore $\mathfrak{p}^f \in \mathcal{P}_p$ and $f(\mathcal{P}_p) = \mathcal{P}_p$. Now assume $f(\mathcal{P}_p) = \mathcal{P}_p$. Let e' be a hyperimaginary given as in Theorem 17.18. Then $f(e') = e'$. Since $e' \in \mathrm{dcl}^{\mathrm{heq}}(a)$, and p does not fork over e', and $p \restriction e'$ is an amalgamation base, we conclude that $e \in \mathrm{dcl}^{\mathrm{heq}}(e')$. Hence $f(e) = e$.

$2 \Rightarrow 1$. By points 1, 2 and 3 of Theorem 17.18. ⊣

COROLLARY 17.20. *In a simple theory every amalgamation base has a canonical base.*

PROOF. By Theorem 17.16 and Corollary 17.19. ⊣

DEFINITION 17.21. If $p(x)$ is an amalgamation base in a simple theory, $\mathrm{Cb}(p)$ is, by definition, $\mathrm{dcl}^{\mathrm{heq}}(e)$ where e is a canonical base of p. For any hyperimaginaries a, b we define $\mathrm{Cb}(a/b) = \mathrm{Cb}(\mathrm{tp}(a/\mathrm{bdd}(b)))$. As mentioned in Remark 17.10, up to interdefinability this notation agrees with the one introduced for canonical bases of stationary types in simple theories.

REMARK 17.22. *Let T be simple. For all hyperimaginaries a, b:*

1. $\mathrm{Cb}(a/b) \subseteq \mathrm{bdd}(b)$.
2. $a \mathop{\smile\!\!\!\!\!\mid}_{\mathrm{Cb}(a/b)} b$.
3. $\mathrm{tp}(a/\mathrm{Cb}(a/b))$ *is an amalgamation base.*
4. *If $d \in \mathrm{bdd}(b)$, $\mathrm{tp}(a/d)$ is an amalgamation base and $a \mathop{\smile\!\!\!\!\!\mid}_d b$, then* $\mathrm{Cb}(a/b) \subseteq \mathrm{dcl}^{\mathrm{heq}}(d)$.

PROOF. By definition of canonical base, since $\mathrm{Cb}(a/b)$ has been defined as $\mathrm{Cb}(\mathrm{tp}(a/\mathrm{bdd}(b)))$. ⊣

LEMMA 17.23. *For simple T the following are equivalent for any hyperimaginaries a, b, c:*

1. $a \mathop{\smile\!\!\!\!\!\!\mid}_b c$.
2. $\mathrm{Cb}(a/bc) = \mathrm{Cb}(a/b)$.
3. $\mathrm{Cb}(a/bc) \subseteq \mathrm{bdd}(b)$.

PROOF. $1 \Rightarrow 2$. If $a \mathop{\smile\!\!\!\!\!\!\mid}_b c$, then $\mathrm{tp}(a/\mathrm{bdd}(bc))$ and $\mathrm{tp}(a/\mathrm{bdd}(b))$ have the same amalgamation class.

$2 \Rightarrow 3$. Since $\mathrm{Cb}(a/b) \subseteq \mathrm{bdd}(b)$.

$3 \Rightarrow 1$. This follows from the fact that $a \mathop{\smile\!\!\!\!\!\!\mid}_{\mathrm{Cb}(a/bc)} bc$. ⊣

PROPOSITION 17.24. *Let T be simple. If $p(x) \in S_E(e)$ is an amalgamation base over a hyperimaginary e and $(a_i : i < \omega)$ is a Morley sequence in p, then $\mathrm{Cb}(p) \subseteq \mathrm{dcl}^{\mathrm{heq}}(a_i : i < \omega)$. Moreover, if T is supersimple and p is of real sort, then $\mathrm{Cb}(p) \subseteq \mathrm{dcl}^{\mathrm{heq}}(a_i : i < n)$ for some $n < \omega$.*

PROOF. Extend the Morley sequence $(a_i : i < \omega)$ to a Morley sequence $(a_i : i \leq \omega)$ in p. By Corollary 17.7 $\mathrm{tp}(a_\omega/(a_i : i < \omega))$ is an amalgamation base. Since $\mathrm{tp}(a_\omega/e(a_i : i < \omega))$ is finitely satisfiable in $\{a_i : i < \omega\}$, by Proposition 17.6 $a_\omega \mathop{\smile\!\!\!\!\!\!\mid}_{(a_i:i<\omega)} e$. Since $a_\omega \mathop{\smile\!\!\!\!\!\!\mid}_e (a_i : i < \omega)$ and $a_\omega \mathop{\smile\!\!\!\!\!\!\mid}_{(a_i:i<\omega)} e$, $p = \mathrm{tp}(a_\omega/e)$ and $\mathrm{tp}(a_\omega/(a_i : i < \omega))$ have the same amalgamation class. Therefore by Theorem 17.18 $\mathrm{Cb}(p) = \mathrm{Cb}(\mathrm{tp}(a_\omega/(a_i : i < \omega))) \subseteq \mathrm{dcl}^{\mathrm{heq}}(a_i : i < \omega)$.

Assume now T is supersimple and $p(x) \in S(e)$ is of real sort. Choose $n < \omega$ such that $a_\omega \mathop{\smile\!\!\!\!\!\!\mid}_{a_{<n}} (a_i : i < \omega)$. As before, $\mathrm{tp}(a_\omega/(a_i : i < \omega))$ is an amalgamation base and $\mathrm{Cb}(p) = \mathrm{Cb}(\mathrm{tp}(a_\omega/(a_i : i < \omega)))$. Therefore $\mathrm{Cb}(p) \subseteq \mathrm{bdd}(a_{<n})$. We claim that $\mathrm{Cb}(p) \subseteq \mathrm{dcl}^{\mathrm{heq}}(a_{\leq n})$. To check it, assume $f \in \mathrm{Aut}(\mathfrak{C}/a_{\leq n})$. Since $a_\omega \overset{\mathrm{Ls}}{\equiv}_{a_{<n}} a_n$, $f(a_\omega) \overset{\mathrm{Ls}}{\equiv}_{a_{<n}} a_n$ and hence $a_\omega \overset{\mathrm{Ls}}{\equiv}_{a_{<n}} f(a_\omega)$. This means that f fixes $\mathrm{tp}(a_\omega/\mathrm{bdd}(a_{<n}))$, which is an amalgamation base with amalgamation class $\mathcal{P}_p$. It follows that f fixes setwise $\mathcal{P}_p$ and hence f fixes $\mathrm{Cb}(p)$. ⊣

Chapter 18

ELIMINATION OF HYPERIMAGINARIES

DEFINITION 18.1. *T eliminates a hyperimaginary e* if there is a sequence of imaginaries $(e_i : i \in I)$ such that $e \sim (e_i : i \in I)$. *T eliminates hyperimaginaries* if T eliminates every hyperimaginary.

PROPOSITION 18.2. *Let $e = a_E$ be a hyperimaginary and let $p(x) = \mathrm{tp}(a)$. Then T eliminates e if and only if there is a family $(E_i : i \in I)$ of 0-definable equivalence relations such that $E \restriction p = (\bigcap_{i \in I} E_i) \restriction p$. In fact it suffices to require that the E_i are 0-definable relations whose restrictions $E_i \restriction p$ to $p(\mathfrak{C})$ are equivalence relations.*

PROOF. It is enough to require that the E_i are equivalence relations on $p(\mathfrak{C})$ since, by compactness, it is always possible to find a formula $\varphi_i(x) \in p$ such that E_i is an equivalence relation on $\varphi_i(\mathfrak{C})$.

Assume $e = a_E$ is a hyperimaginary and choose a family $(E_i : i \in I)$ of 0-definable equivalence relations such that on $p(x) = \mathrm{tp}(a)$ the equivalence relation E agrees with $\bigcap_{i \in I} E_i$. Then $e \sim (e_i : i \in I)$ where $e_i = {a_i}_{E_i}$, if a_i is the subtuple of a corresponding to the variables of E_i. For the other direction, by assumption there is a sequence $(e_i : i \in I)$ of imaginaries $e_i = {a_i}_{E_i}$ such that $e \sim (e_i : i \in I)$. Let $p_i(x, y) = \mathrm{tp}(aa_i)$ for each $i \in I$. We may assume the family is closed under finite concatenation in the following sense: for each $n < \omega$, for all $i_0, \dots, i_n \in I$ there is some $j \in I$ such that $(e_{i_0}, \dots, e_{i_n}) \sim e_j$. Therefore $b_1 \equiv_e b_2$ if and only if $b_1 \equiv_{e_i} b_2$ for all $i \in I$. Then:

1. $E(x, x') \cup p_i(x, y) \cup p_i(x', y') \vdash E_i(y, y')$,
2. $\models p_i(a, a_i)$,
3. $p(x) \vdash \exists y p_i(x, y)$.

By compactness we can substitute a single formula $\varphi_i(x, y) \in p_i$ for $p_i(x, y)$ and still have property 1. We then define for each $i \in I$, $\psi(x, y) \in p_i(x, y)$:

$$F_{i,\psi}(y, z) \ \Leftrightarrow\ \exists uv(E_i(u, v) \wedge \varphi_i(y, u) \wedge \psi(y, u) \wedge \varphi_i(z, v) \wedge \psi(z, v)).$$

Clearly $F_{i,\psi}$ is definable over $\emptyset$ and $F_{i,\psi} \restriction p$ is an equivalence relation. Moreover $E \restriction p = (\bigcap_{i \in I, \psi \in p_i} F_{i,\psi}) \restriction p$. The only point that needs to be checked is that $E(b_1, b_2)$ if we assume $F_{i,\psi}(b_1, b_2)$ for all i, ψ. It is enough to check it for the case $a = b_1$. Fix some $i \in I$. By compactness, there are c_1, c_2 such that $\models p_i(a, c_1)$, $\models p_i(b_2, c_2)$ and $E_i(c_1, c_2)$. Since $\models p_i(a, a_i)$, $E_i(a_i, c_1)$ and therefore $E_i(a_i, c_2)$. There is some automorphism f such that $f(b_2c_2) = aa_i$. Since f fixes e_i, $a \equiv_{e_i} b_2$. Hence $a \equiv_e b_2$, which implies $E(a, b_2)$. ⊣

COROLLARY 18.3. *T eliminates hyperimaginaries if and only if for any $p(x) \in S(\emptyset)$, for any 0-type-definable equivalence relation E on $p(\mathfrak{C})$, there is a family $(E_i : i \in I)$ of 0-definable equivalence relations such that $E = (\bigcap_{i\in I} E_i) \restriction p$. In fact it suffices to require that the E_i are 0-definable relations whose restrictions $E_i \restriction p$ to $p(\mathfrak{C})$ are equivalence relations.*

PROOF. By Proposition 18.2. ⊣

LEMMA 18.4. *Let E be an intersection of definable (possibly with parameters) equivalence relations. If E is type-definable over $\emptyset$, then E is an intersection of 0-definable equivalence relations.*

PROOF. Let $E = \bigcap_{i\in I} E_i$ where every E_i is an equivalence relation, defined by $\varphi_i(x, y, a_i)$, with $\varphi_i(x, y, z) \in L$. Assume $\Sigma(x, y)$ is a type over $\emptyset$ defining E and let $p_i(z) = \mathrm{tp}(a_i)$. Then $\Sigma(x, y) \cup p_i(z) \vdash \varphi_i(x, y, z)$, and therefore $\Sigma(x, y) \vdash \forall z\,(\psi_i(z) \to \varphi_i(x, y, z))$ for some $\psi_i(z) \in p_i$. We can choose it so that $\psi_i(z)$ implies that $\varphi_i(x, y, z)$ is an equivalence relation in x, y. Then

$$\forall z\,(\psi_i(z) \to \varphi_i(x, y, z))$$

defines (over $\emptyset$) an equivalence relation F_i such that $E \subseteq F_i \subseteq E_i$. Hence $E = \bigcap_{i\in I} F_i$. ⊣

PROPOSITION 18.5. *If T eliminates hyperimaginaries, then $T(A)$ also eliminates hyperimaginaries.*

PROOF. By Lemma 15.20. ⊣

LEMMA 18.6. *If $e \in \mathrm{dcl}^{\mathrm{heq}}(a)$ for some sequence of imaginaries $a \in \mathrm{bdd}(e)$, then e is eliminable.*

PROOF. Let $a = (a_i : i \in I)$ where every a_i is an imaginary. For each finite $J \subseteq I$ let $a_J = (a_i : i \in J)$. Then $a_J \in \mathrm{acl}^{\mathrm{heq}}(e) \cap \mathfrak{C}^{\mathrm{eq}}$. Consider the finite set $\{f(a_J) : f \in \mathrm{Aut}(\mathfrak{C}/e)\}$ as a single imaginary b_J, as explained in Corollary 1.12. Now let $b = (b_J : J \subseteq I \text{ is finite})$. It is clear that $b \in \mathrm{dcl}^{\mathrm{heq}}(e)$. We check now that $e \in \mathrm{dcl}^{\mathrm{heq}}(b)$. Assume f fixes b. Then for each finite J, $a_J \equiv_e f(a_J)$ and therefore $a \equiv_e f(a)$. Hence $f(a)e \equiv ae \equiv f(a)f(e)$. Since $e \in \mathrm{dcl}^{\mathrm{heq}}(a)$, also $f(e) \in \mathrm{dcl}^{\mathrm{heq}}(f(a))$ and hence $e = f(e)$. ⊣

PROPOSITION 18.7. *Let T be simple. If $e = a_E$ is a hyperimaginary, then $e \in \mathrm{Cb}(a/e)$.*

PROOF. In a first step we show that $e \in \mathrm{bdd}(\mathrm{Cb}(a/e))$. By Remark 17.22, $a \mathop{\smile\!\!\!\!\!\!|}_{\mathrm{Cb}(a/e)} e$ and since $e \in \mathrm{dcl}^{\mathrm{heq}}(a)$, we see that $e \mathop{\smile\!\!\!\!\!\!|}_{\mathrm{Cb}(a/e)} e$. By Proposition 16.19, $e \in \mathrm{bdd}(\mathrm{Cb}(a/e))$. Now we show that e is definable over $\mathrm{Cb}(a/e)$. Since $\mathrm{tp}(a/\mathrm{Cb}(a/e))$ is an amalgamation base, by Remark 17.2

$\text{tp}(a/\text{Cb}(a/e)) = \text{tp}(a/\text{bdd}(\text{Cb}(a/e))) \vdash \text{tp}(a/e)$. Let $f \in \text{Aut}(\mathfrak{C}/\text{Cb}(a/e))$. Then $a \equiv_{\text{Cb}(a/e)} f(a)$ and therefore $a \equiv_e f(a)$. It follows that $E(a, f(a))$ and hence $f(e) = e$. ⊣

LEMMA 18.8. *Let T be simple. If $a = (a_i : i \in I)$ is a sequence of real elements and for all finite $J \subseteq I$, $a_J = (a_i : i \in J)$, then for every hyperimaginary e, $\text{Cb}(a_J/e) \subseteq \text{Cb}(a_{J'}/e)$ for $J \subseteq J'$ and*

$$\text{Cb}(a/e) = \text{dcl}^{\text{heq}}\Big(\bigcup_{J \subseteq I \text{ finite}} \text{Cb}(a_J/e)\Big).$$

PROOF. Let C be the union of all $\text{Cb}(a_J/e)$ for $J \subseteq I$ finite. Since $a_J \mathop{\smile\!\!\!\!\!\!|}_{\text{Cb}(a/e)} e$ and (by Remark 17.4) $\text{tp}(a_J/\text{Cb}(a/e))$ is an amalgamation base, by point 4 of remark 17.22, $\text{Cb}(a_J/e) \subseteq \text{Cb}(a/e)$. Similarly, one can check that $\text{Cb}(a_J/e) \subseteq \text{Cb}(a_{J'}/e)$ if $J \subseteq J'$. We claim that $\text{tp}(a/C)$ is an amalgamation base. Since $a \mathop{\smile\!\!\!\!\!\!|}_C e$, this will give $\text{Cb}(a/e) \subseteq \text{dcl}^{\text{heq}}(C)$. We use Remark 17.2. By Corollary 16.36 it is enough to show that $\text{tp}(a/C) \vdash \text{tp}(a/\text{bdd}(\text{Cb}(a_J/e))$ for all finite J, and this is clear since $\text{tp}(a_J/\text{Cb}(a_J/e))$ is an amalgamation base and therefore

$$\text{tp}(a_J/\text{Cb}(a_J/e)) \vdash \text{tp}(a_J/\text{bdd}(\text{Cb}(a_J/e))). \qquad ⊣$$

PROPOSITION 18.9. *Let T be simple. If for each finitary amalgamation base $p(x)$, the canonical base $\text{Cb}(p)$ is eliminable, then T eliminates hyperimaginaries.*

PROOF. Let $e = a_E$ be a hyperimaginary. By Lemma 18.8 and our assumption, there is a sequence d of imaginaries such that $\text{Cb}(a/e) \sim d$. Then $d \in \text{bdd}(e)$. By Proposition 18.7

$$e \in \text{Cb}(a/e) \subseteq \text{dcl}^{\text{heq}}(d).$$

By Lemma 18.6 $e \sim d'$ for some sequence of imaginaries d'. ⊣

COROLLARY 18.10. *Stable theories eliminate hyperimaginaries.*

PROOF. By Proposition 18.9 and by the fact that canonical bases in stable theories are sequences of imaginaries (see Proposition 17.3). ⊣

COROLLARY 18.11. *If a supersimple theory eliminates all finitary hyperimaginaries, then it eliminates all hyperimaginaries.*

PROOF. By Proposition 18.9 it suffices to show that T eliminates $\text{Cb}(p)$ for all finitary amalgamation bases $p(x)$. This follows from the assumption since by Proposition 17.24 $\text{Cb}(p)$ is definable over a finite set and therefore it is finitary. ⊣

PROPOSITION 18.12. *Let E be a 0-type-definable equivalence relation and let E^* be the equivalence relation given by*

$$E^*(a,b) \Leftrightarrow E(a',b') \text{ for some } a' \equiv a,\ b' \equiv b.$$

Then E is an intersection of 0*-definable equivalence relations if and only if E^* is an intersection of* 0*-definable equivalence relations and for each $p(x) \in S(\emptyset)$, $E \restriction p = E \cap (p(\mathfrak{C}) \times p(\mathfrak{C}))$ is an intersection of* 0*-definable equivalence relations.*

PROOF. Note that each of the two following conditions is equivalent to $E^*(a, b)$:

1. $E(a, c)$ for some $c \equiv b$.
2. $E(a, a')$ and $E(b, b')$ for some $a' \equiv b'$.

If $E^*(a, b)$ is witnessed by $E(a', b')$ where $a' \equiv a$ and $b' \equiv b$, and we choose c such that $ac \equiv a'b'$ then c witnesses that 1 holds. For $1 \Rightarrow 2$ just take $a' = c$ and $b' = b$. Finally if a', b' are as in 2 and we choose c such that $a'a \equiv b'c$, then $E(c, b)$, $a \equiv c$ and $b \equiv b$.

Now assume $E = \bigcap_{i \in I} E_i$ for 0-definable equivalence relations E_i. Then obviously for each $p \in S(\emptyset)$, E agrees with $\bigcap_{i \in I} E_i$ on p. Moreover $E^* = \bigcap_{i \in I, \varphi \in L} E_{i\varphi}$ where $E_{i\varphi}(x, y)$ is the equivalence relation defined by

$$\exists z(\varphi(z) \wedge E_i(x, z)) \leftrightarrow \exists z(\varphi(z) \wedge E_i(y, z)).$$

For the other direction, suppose that E^* is an intersection of 0-definable equivalence relations and for each $p(x) \in S(\emptyset)$, $E \restriction p = \bigcap_{i \in I_p} E_{ip} \restriction p$ for a family of 0-definable equivalence relations E_{ip}. We can assume that the type $E(x, y)$ defining the equivalence relation E is made of symmetric formulas. Fix some $p(x) \in S(\emptyset)$ and choose some $a \models p$. For each $i \in I_p$ we can find some formula $\sigma_{ip}(x, y) \in E(x, y)$ and some $\psi_{ip}(x) \in p$ such that

$$\sigma_{ip}(x, y) \wedge \psi_{ip}(x) \wedge \psi_{ip}(y) \vdash E_{ip}(x, y).$$

We can also find some $\overline{\sigma}_{ip}(x, y) \in E(x, y)$ such that

$$\overline{\sigma}_{ip}(x, y) \wedge \overline{\sigma}_{ip}(y, z) \wedge \overline{\sigma}_{ip}(z, u) \vdash \sigma_{ip}(x, u)$$

and some 0-definable equivalence relation E^*_{ip} in the family whose intersection is E^* such that

$$E^*_{ip}(x, a) \vdash \exists y(\psi_{ip}(y) \wedge \overline{\sigma}_{ip}(x, y)).$$

Consider the relation $F_{ip}(x, y)$ defined by the disjunction of $(\neg E^*_{ip}(x, a) \wedge \neg E^*_{ip}(y, a))$ with

$$(E^*_{ip}(x, a) \wedge E^*_{ip}(y, a) \wedge \exists uv(\psi_{ip}(u) \wedge \psi_{ip}(v) \wedge \overline{\sigma}_{ip}(x, u) \wedge \overline{\sigma}_{ip}(y, v) \wedge E_{ip}(u, v))).$$

Note that the definition is in fact independent of the choice of the realization a of p. It is clearly reflexive and symmetric. It is not difficult to see that it is also transitive. We claim that

$$E = E^* \cap \bigcap_{p \in S(\emptyset), i \in I_p} F_{ip}.$$

By Lemma 18.4 this will show that E is an intersection of 0-definable equivalence relations.

Assume $E(c,d)$. Then $E^*(c,d)$. Let $p(x) \in S(\emptyset)$, let $i \in I_p$, and let $a \models p$. We want to check that $F_{ip}(c,d)$. We may assume $E^*_{ip}(c,a) \wedge E^*_{ip}(d,a)$. By choice of E^*_{ip} we know that there are c', d' such that $\models \psi_{ip}(c') \wedge \overline{\sigma}_{ip}(c,c')$ and $\models \psi_{ip}(d') \wedge \overline{\sigma}_{ip}(d,d')$. Then $\models \sigma_{ip}(c',d')$ and therefore $E_{ip}(c',d')$.

For the other direction, assume $E^*(c,d)$ and $F_{ip}(c,d)$ for all p,i. Let $p(x) = \mathrm{tp}(c)$. As remarked above, $E(c',d)$ for some $c' \equiv c$. It is enough to show that $E(c,c')$ and for this we have to check that $E_{ip}(c,c')$ for all $i \in I_p$. Note that $F_{ip}(c,d)$ and $F_{ip}(d,c')$ since we have already shown that $E(x,y)$ implies $F_{ip}(x,y)$. Hence $F_{ip}(c,c')$ and by definition of F_{ip} there are b,b' such that $\models \psi_{ip}(b) \wedge \overline{\sigma}_{ip}(c,b) \wedge \psi_{ip}(b') \wedge \overline{\sigma}_{ip}(c',b') \wedge E_{ip}(b,b')$. Note that $\models \psi_{ip}(c) \wedge \psi_{ip}(b) \wedge \sigma_{ip}(c,b)$ and thus $E_{ip}(c,b)$. Similarly $E_{ip}(c',b')$ and we conclude $E_{ip}(c,c')$. ⊣

LEMMA 18.13. *Let T be small and let E be a 0-type-definable equivalence relation on $\mathfrak{C}^n$ such that: if $E(a,b)$, $a \equiv a'$ and $b \equiv b'$, then $E(a',b')$. Then E is an intersection of 0-definable equivalence relations.*

PROOF. We claim that whenever $\neg E(a,b)$, then for some $\varphi_{ab} \in L$, $\models \varphi_{ab}(a) \wedge \neg\varphi_{ab}(b)$ and $E(x,y) \wedge \varphi_{ab}(x) \wedge \neg\varphi_{ab}(y)$ is inconsistent. If this is the case, we can then express E as an intersection of 0-definable equivalence relations as follows:

$$E(x,y) \Leftrightarrow \bigwedge_{\neg E(a,b)} \varphi_{ab}(x) \leftrightarrow \varphi_{ab}(y).$$

In order to prove this claim, assume $\neg E(a,b)$ and set $p(x) = \mathrm{tp}(a)$, $q(x) = \mathrm{tp}(b)$. Recall that, by Remark 14.27, all types in $S_n(\emptyset)$ have ordinal Cantor–Bendixson rank. We first observe that $E(x,y) \cup p(x) \cup q(y)$ is inconsistent and hence we can choose $\varphi(x) \in p(x)$, $\psi(y) \in q(y)$ such that $E(x,y) \wedge \varphi(x) \wedge \psi(y)$ is inconsistent and $\neg\varphi(x) \wedge \neg\psi(x)$ is of minimal Cantor–Bendixson rank α in the space $S_n(\emptyset)$ and of minimal degree in this rank. If $\neg\varphi(x) \wedge \neg\psi(x)$ is inconsistent we set $\varphi_{ab} = \varphi$ and this choice satisfies the requirements. Otherwise we choose a type $p'(x) \in S_n(\emptyset)$ of rank α containing the formula $\neg\varphi(x) \wedge \neg\psi(x)$ and also a realization $c \models p'$. Now, if there if some $a' \models \varphi$ and some $b' \models \psi$ such that $E(a',c)$ and $E(b',c)$ then $E(x,y) \wedge \varphi(x) \wedge \psi(y)$ turns out to be consistent. Hence we may assume that there is no $a' \models \varphi$ such that $E(a',c)$, that is, $E(x,y) \wedge \varphi(x) \wedge p'(y)$ is inconsistent. Therefore $E(x,y) \wedge \varphi(x) \wedge \psi'(y)$ is inconsistent for some $\psi' \in p'$. Note that either $\neg\varphi(x) \wedge \neg\psi'(x)$ has rank $< \alpha$ or has rank α and smaller degree than $\neg\varphi(x) \wedge \neg\psi(x)$. This contradicts the previous choice of $\varphi(x)$ and $\psi(x)$. ⊣

THEOREM 18.14. *Let T be small.*

1. *If E is a 0-type-definable equivalence relation on $\mathfrak{C}^n$, then E is an intersection of 0-definable equivalence relations.*
2. *T eliminates all finitary hyperimaginaries.*
3. *For any finite set A, $\overset{\mathrm{KP}}{\equiv}_A = \overset{\mathrm{s}}{\equiv}_A$.*

4. *If T is G-compact over A, in particular if T is simple, then* $\overset{\text{Ls}}{\equiv}_A = \overset{\text{s}}{\equiv}_A$.

PROOF. 1. We apply Proposition 18.12. It is clear that E^* satisfies the hypothesis of Lemma 18.13 and therefore it is an intersection of 0-definable equivalence relations. Now fix some $p(x) \in S_n(\emptyset)$ and choose some $c \models p$. We have to show that for some family $(E_i : i \in I)$ of 0-definable equivalence relations, $E \restriction p = \bigcap_{i \in I} E_i \restriction p$. Consider the relation

$$F(x, y) \Leftrightarrow \text{ for some } z,\ cx \equiv zy \text{ and } E(c, z).$$

It is an equivalence relation and it is type-definable over c. Since $T(c)$ is small and in $T(c)$ the relation F satisfies the hypothesis of Lemma 18.13, there is some family $(F_i : i \in I)$ of equivalence relations F_i such that $F = \bigcap_{i \in I} F_i$ and for each $i \in I$ there is some $\varphi_i(x, y, z) \in L$ such that $\varphi_i(x, y, c)$ defines F_i. Now let

$$E_i(x, y) \Leftrightarrow \forall u(\varphi_i(u, x, x) \leftrightarrow \varphi_i(u, y, y)).$$

It is clearly an equivalence relation. We check that $E \restriction p = \bigcap_{i \in I} E_i \restriction p$. It suffices to see that for any $a \models p$, $E(a, c)$ if and only if $E_i(a, c)$ for all $i \in I$. Assume $E(a, c)$, let $i \in I$, let b be arbitrary and choose b' such that $ab \equiv cb'$. Then $F(b, b')$ and therefore $\models \varphi_i(b, b', c)$. Since $\varphi_i(x, y, c)$ defines an equivalence relation, $\models \varphi_i(b, c, c) \leftrightarrow \varphi_i(b', c, c)$. By automorphism, $\models \varphi_i(b, c, c) \leftrightarrow \varphi_i(b, a, a)$ and thus $E_i(a, c)$. For the other direction, assume $E_i(a, c)$ for all $i \in I$. Since $\models \varphi_i(a, a, a)$, we obtain $\models \varphi_i(a, c, c)$ and hence $F(a, c)$. This clearly implies $E(a, c)$.

2. It follows from 1 and Corollary 18.3.

3. Since $T(A)$ is again small we may assume $A = \emptyset$. It is enough to check the equality for finite sequences and this case follows straightforwardly from 1 since it implies that on n-tuples $\overset{\text{KP}}{\equiv}$ is an intersection of finite 0-definable equivalence relations.

4. If T is simple, then by Corollary 10.17 T is G-compact over A, that is, $\overset{\text{KP}}{\equiv}_A = \overset{\text{Ls}}{\equiv}_A$. By Remark 15.22, $a \overset{\text{KP}}{\equiv}_A b$ if and only if $a \overset{\text{KP}}{\equiv}_{A'} b$ for all finite $A' \subseteq A$. The same for $\overset{\text{s}}{\equiv}_A$. Then we can apply 3. ⊣

EXAMPLE 18.15. 1. (Pillay–Poizat in [35]) There is a superstable theory T (of U-rank 1) where we can find a 0-type-definable equivalence relation which is not an intersection of 0-definable equivalence relations.

2. (Adler in [1]) There is an ω-categorical (hence small) theory which does not eliminate hyperimaginaries. More generally, in any theory with the strict order property there is an infinitary hyperimaginary which is not eliminable.

DEFINITION 18.16. A formula $\varphi(x, y) \in L$ is *low* if there is some $n < \omega$ such that for any indiscernible sequence $(a_i : i < \omega)$, if $\{\varphi(x, a_i) : i < \omega\}$ is inconsistent, then it is n-inconsistent. We say that T is *low* if it is simple and every formula is low in T.

DEFINITION 18.17. Let $\varphi(x, y) \in L$. For any set of formulas $\pi(x)$ we define the rank $\mathrm{D}(\pi, \varphi)$ as follows

1. $\mathrm{D}(\pi, \varphi) \geq 0$ if and only if $\pi(x)$ is consistent.
2. $\mathrm{D}(\pi, \varphi) \geq \alpha + 1$ if and only if for some a, $\varphi(x, a)$ divides over the parameters of π and $\mathrm{D}(\pi \cup \{\varphi(x, a)\}, \varphi) \geq \alpha$.
3. $\mathrm{D}(\pi, \varphi) \geq \alpha$ if and only if $\mathrm{D}(\pi, \varphi) \geq \beta$ for all $\beta < \alpha$ if α is a limit ordinal number.

REMARK 18.18. 1. $\mathrm{D}(\pi, \varphi, k) \leq \mathrm{D}(\pi, \varphi) \leq \mathrm{D}(\pi)$.
2. *If $\pi(x)$ is a partial type over A, then $\mathrm{D}(\pi, \varphi) \geq \alpha + 1$ if and only if for some a, $\varphi(x, a)$ divides over A and $\mathrm{D}(\pi \cup \{\varphi(x, a)\}, \varphi) \geq \alpha$.*

PROPOSITION 18.19. *Let T be simple and let $\varphi(x, y) \in L$. The following are equivalent:*

1. *$\varphi(x, y)$ is low.*
2. *There is some $k < \omega$ such that for all π, $\mathrm{D}(\pi, \varphi) = \mathrm{D}(\pi, \varphi, k)$.*
3. *$\mathrm{D}(x = x, \varphi) < \omega$.*
4. *There is some $n < \omega$ such that for all $k < \omega$, $\mathrm{D}(x = x, \varphi, k) < n$.*
5. *There is some $n < \omega$ such that φ divides at most n times.*
6. *$\{(a, b) \in \mathfrak{C} : \varphi(x, a)$ divides over $b\}$ is type-definable over $\emptyset$ (for any fixed length of b).*

PROOF. $1 \Rightarrow 2$. Fix $n < \omega$ as in the definition of low. If $\varphi(x, a)$ divides over A, it divides over A with respect to n. Hence $\mathrm{D}(\pi, \varphi) = \mathrm{D}(\pi, \varphi, n)$.

$2 \Rightarrow 3$. By simplicity, $\mathrm{D}(x = x, \varphi, k) < \omega$.

$3 \Rightarrow 4$ is clear since $\mathrm{D}(x = x, \varphi, k) \leq \mathrm{D}(x = x, \varphi)$.

$4 \Rightarrow 5$. Fix n as in 4. If φ divide m times, there are tuples $(a_i : i < m)$ and numbers $(k_i : i < m)$ such that $\{\varphi(x, a_i) : i < m\}$ is consistent and for each $i < m$, $\varphi(x, a_i)$ divides over $a_{<i}$ with respect to k_i. If $k = \max_{i<m} k_i$ then $\varphi(x, a_i)$ divides over $a_{<i}$ with respect to k and hence $m \leq n$.

$5 \Rightarrow 1$. If $(a_i : i < \omega)$ is indiscernible, and $\{\varphi(x, a_i) : i < \omega\}$ is inconsistent but not k-inconsistent, then $(a_i : i < k)$ witnesses that $\varphi(x, y)$ divides k times.

$1 \Rightarrow 6$. It follows from 1 that there is some $k < \omega$ such that for all sequences a, b if $\varphi(x, a)$ divides over b, then $\varphi(x, a)$ divides over b with respect to k. But $\{(a, b) \in \mathfrak{C} : \varphi(x, a)$ divides over b with respect to $k\}$ is type-definable over $\emptyset$.

$6 \Rightarrow 1$. Assume that $\varphi(x, y)$ is not low. For each $k < \omega$ let $(a_i^k : i < \omega + \omega)$ be an indiscernible sequence such that $\{\varphi(x, a_i^k) : i < \omega + \omega\}$ is inconsistent but not k-inconsistent. Choose a nonprincipal ultrafilter D over ω and let $(c_i : i < \omega + \omega)$ be a realization of the ultraproduct of types $\prod_{\mathrm{D}}(p_k : k < \omega)$ where $p_k = \mathrm{tp}(a_i^k : i < \omega + \omega)$. Then $(c_i : i < \omega + \omega)$ is indiscernible and $\{\varphi(x, c_i) : \omega \leq i < \omega + \omega\}$ is k-consistent for every $k < \omega$ and hence it is consistent. Let $c = (c_i : i < \omega)$. By Lemma 10.4, $(c_i : \omega \leq i < \omega + \omega)$ is a Morley sequence over c. By Proposition 5.15 $\varphi(x, c_\omega)$ does not divide over c.

Assume $\pi(x, y)$ is a partial type over $\emptyset$ defining dividing as in 6. Since, for each $k < \omega$, $\varphi(x, a^k_\omega)$ divides over $b_k = (a^k_i : i < \omega)$, we have $\models \pi(a^k_\omega, b_k)$ and therefore $\models \pi(c_\omega, c)$. But then $\varphi(x, c_\omega)$ divides over c. ⊣

REMARK 18.20. *Conditions* 3, 4, 5 *of Proposition* 18.19 *are equivalent in any theory* T. *Moreover, if they hold for any* φ, *the theory* T *is simple.*

PROOF. Note that $5 \Rightarrow 3$ follows from Proposition 3.8. ⊣

REMARK 18.21. *If* T *is low, then* $T(A)$ *is also low for any set* A.

PROOF. This is clear, for instance, from point 6 of Proposition 18.19 since $\varphi(x, a, b, c)$ divides over bc in T if and only if $\varphi(x, a, b, c)$ divides over b in $T(c)$. ⊣

PROPOSITION 18.22. 1. *Any stable theory is low.*
2. *Any supersimple theory of finite* D*-rank is low.*

PROOF. By Proposition 18.19, 2 is clear, since $\mathrm{D}(x = x, \varphi) \leq \mathrm{D}(x = x)$. For 1, assume $\{\varphi(x, a_i) : i < n\}$ is consistent and $\varphi(x, a_i)$ divides over $(a_j : j < i)$ for each $i < n$. Let $b \models \bigwedge_{i<n} \varphi(x, a_i)$ and let $p_i(x) = \mathrm{tp}(b/\{a_j : j < i\})$ for $i = 1, \dots, n$. By Corollary 8.8, $\mathrm{CB}_\varphi(p_i) > \mathrm{CB}_\varphi(p_{i+1})$ for all $i < n$ and therefore $\omega > \mathrm{CB}_\varphi(x = x) \geq \mathrm{D}(x = x, \varphi)$. ⊣

REMARK 18.23. *There are supersimple nonlow theories. An example is given in* [9].

DEFINITION 18.24. There is a natural topology in $\mathfrak{C}^I$, the topology whose closed sets are the type-definable (with parameters) subsets of $\mathfrak{C}^I$. By analogy with the case of an algebraically closed field, let us call it the Zariski topology. If E is a 0-type-definable equivalence relation in a type-definable over $\emptyset$ subclass of $\mathfrak{C}^I$, the *logic topology* or the *Kim–Pillay topology* is the quotient topology of the Zariski topology. If $\pi(x)$ is the type defining the domain of E and $X = \pi(\mathfrak{C})/E$ is the quotient, then $A \subseteq X$ is closed if and only if $\{a \models \pi : a_E \in A\}$ is type-definable. As usual, in this context we will always identify E with the type defining it and we will assume that the type $E(x, y)$ is closed under finite conjunctions and that $E(x, y) \vdash \pi(x) \cup \pi(y)$.

PROPOSITION 18.25. *Let* $\pi(x)$ *be a type over* $\emptyset$, *let* E *be a* 0*-type-definable equivalence relation on* $\pi(\mathfrak{C})$, *and let us work in the Kim–Pillay space* $X = \pi(\mathfrak{C})/E$.

1. $Y \subseteq X$ *is closed if and only if for some type-definable class* A, $Y = \{a_E : a \in A\}$.
2. X *is Hausdorff.*
3. *A basis of open sets is given by the collection of all*
$$U_{a\varphi} = \{b_E : \models \varphi(a', b') \textit{ for all } a', b' \textit{ such that } E(a, a'), E(b, b')\}$$
where $a \models \pi$ *and* $\varphi = \varphi(x, y) \in E$.
4. X *is compact if and only if* E *is bounded.*

PROOF. 1. If $A = \Phi(\mathfrak{C})$, then $\{a : a_E \subset Y\}$ is defined by the type

$$\Psi(x) = \exists y(E(x,y) \wedge \Phi(y)).$$

2 is clear. We check 3. Note that $\{b \models \pi : b_E \not\in U_{a\varphi}\}$ is type-definable and hence $U_{a\varphi}$ is open. Let U be open and $a_E \in U$. Choose a partial type $\Sigma(x)$ (extending π) such that $X \setminus U$ is the set of all b_E such that $b \models \Sigma$. Choose $\sigma \in \Sigma$ such that $\models \neg\sigma(a)$. Note that $E(x,y) \wedge \Sigma(x) \vdash \Sigma(y)$ and hence $\varphi(x,y) \wedge \Sigma(x) \vdash \sigma(y)$ for some $\varphi(x,y) \in E(x,y)$. Then $a_E \in U_{a\varphi} \subseteq U$.

4. Assume first E is bounded and let $(F_i : i \in I)$ be a family of closed sets with the finite intersection property. For each $i \in I$ choose a type $\Phi_i(x)$ such that $F_i = \{a_E : a \models \Phi_i\}$. If the number of E-classes is bounded by κ, the number of closed sets in X is bounded by 2^κ and hence $|I| \leq 2^\kappa$. Therefore $\bigcup_{i \in I} \Phi_i$ is a partial type over a subset of $\mathfrak{C}$ and we can realize it by some $a \in \mathfrak{C}$. Clearly, $a_E \in F_i$ for all $i \in I$. For the other direction, assume now X is compact. Fix $\varphi(x,y) \in E(x,y)$. We will show that φ is finite on π, that is, there is no infinite sequence $(a_i : i \in \omega)$ of realizations a_i of π such that $\models \neg\varphi(a_i, a_j)$ for all $i < j < \omega$. From this it follows that E is bounded. Assume there is such a sequence $(a_i : i \in \omega)$. We can extend it to a maximal one $(a_i : i \in I)$. Then for any $a \models \pi$ there is some $i \in I$ such that $\models \varphi(a, a_i)$, that is $X \subseteq \bigcup_{i \in I} U_{a_i\varphi}$. By compactness of X, for some finite $I_0 \subseteq I$, $X \subseteq \bigcup_{i \in I_0} U_{a_i\varphi}$. This contradicts the choice of the sequence. ⊣

PROPOSITION 18.26. *Let $\pi(x)$ be a type over $\emptyset$, let E be a 0-type-definable equivalence relation on $\pi(\mathfrak{C})$, and consider the Kim–Pillay space $X = \pi(\mathfrak{C})/E$. The following conditions are equivalent.*

1. *X is 0-dimensional.*
2. *E is an intersection of 0-definable equivalence relations.*
3. *For each $\varphi(x,y) \in E$ there is some $\varphi'(x,y)$ implied by $E(x,y)$ and such that*
   (a) $\pi(x) \cup \pi(y) \vdash \varphi'(x,y) \to \varphi(x,y)$,
   (b) $E(x,x') \cup E(y,y') \vdash \varphi'(x,y) \to \varphi'(x',y')$.

PROOF. $1 \Rightarrow 2$. Let $(O_i : i \in I)$ be a basis of clopen sets. For each $i \in I$ there is some formula $\varphi_i(x) \in L(\mathfrak{C})$ such that $\{a \models \pi : \models \varphi_i(a)\} = \{a \models \pi : a_E \in O_i\}$. Let $a \models \pi$. Since $\{a_E\}$ is closed, there is a subset $I_a \subseteq I$ such that $\{a_E\} = \bigcap_{i \in I_a} O_i$. For each $i \in I$, $(\varphi_i(x) \leftrightarrow \varphi_i(y))$ defines an equivalence relation. It is easy to check that E can be defined by

$$\bigwedge_{a \models \pi} \bigwedge_{i \in I_a} (\varphi_i(x) \leftrightarrow \varphi_i(y)).$$

$2 \Rightarrow 3$. Let $E = \bigcap_{i \in I} E_i$ where each E_i is a 0-definable equivalence relation. If $\varphi(x,y) \in E(x,y)$, then for some $i \in I$, $E_i(x,y) \vdash \varphi(x,y)$ and clearly $\varphi'(x,y) = E_i(x,y)$ satisfies all the requirements.

$3 \Rightarrow 1$. Let $(U_{a\varphi} : a \models \pi,\ \varphi \in E)$ be the basis of open sets described in Proposition 18.25. For each $\varphi \in E$ choose φ' as in 3. Then $(U_{a\varphi'} : a \models \pi,\ \varphi \in E)$ is again a basis of open sets. It is easy to check that in fact each $U_{a\varphi'}$ is clopen. ⊣

PROPOSITION 18.27. *If T is G-compact over $\emptyset$ (in particular, if T is simple), then T eliminates all bounded hyperimaginaries if and only if* Lstp = stp, *that is, if and only if for all sequences a, b: $a \overset{\text{Ls}}{\equiv} b$ if and only if $a \overset{\text{s}}{\equiv} b$.*

PROOF. G-compactness over $\emptyset$ means $\text{Autf}(\mathfrak{C}) = \text{Aut}(\mathfrak{C}/\text{bdd}(\emptyset))$ and also means $\overset{\text{KP}}{\equiv} = \overset{\text{Ls}}{\equiv}$. It is clear that if T eliminates all bounded hyperimaginaries, then $\text{Aut}(\mathfrak{C}/\text{bdd}(\emptyset)) = \text{Aut}(\mathfrak{C}/\text{acl}^{\text{eq}}(\emptyset))$ and therefore Lstp = stp. For the other direction, let $e = a_E$ be a bounded hyperimaginary. By Proposition 15.27 we can assume E is a bounded equivalence relation. By G-compactness, $\overset{\text{Ls}}{\equiv}$ is the least bounded 0-type-definable equivalence relation and therefore a_E splits into a bounded number of Lascar strong types. By assumption and by Proposition 18.2 for each bEa there is a sequence of imaginaries b' such that $b_{\overset{\text{Ls}}{\equiv}} \sim b'$. Let $(b_i : i \in I)$ be a sequence of representatives of Lascar strong types of elements in a_E. Then $e \in \text{dcl}^{\text{heq}}(b'_i : i \in I)$ and $(b'_i : i \in I) \in \text{bdd}(e)$. By Lemma 18.6 e is equivalent to a sequence of imaginaries. ⊣

LEMMA 18.28. *Assume T is a simple theory. Let $p(y) \in S(A)$ and let $\psi_1(x, y), \dots, \psi_n(x, y) \in L(A)$. Then*

$$\{(a_1, \dots, a_n) : a_1, \dots, a_n \text{ are } A\text{-independent realizations of } p \text{ and the formula } \psi_1(x, a_1) \wedge \cdots \wedge \psi_n(x, a_n) \text{ does not fork over } A\}$$

is type-definable over A.

PROOF. By Proposition 5.15 and Corollary 5.24, noticing that if each $b_i = (a_i^1, \dots, a_i^n)$ is an A-independent sequence of realizations a_i^j of p, then $(b_i : i < \omega)$ is a Morley sequence over A if and only if the composed sequence $(c_k : k < \omega)$ (where $c_{i \cdot n + j} = a_i^j$) is a Morley sequence over A. ⊣

THEOREM 18.29. *Let T be a low theory.*

1. *T eliminates all bounded hyperimaginaries.*
2. *For any set A, for all tuples a, b: $a \overset{\text{s}}{\equiv}_A b$ if and only if $a \overset{\text{Ls}}{\equiv}_A b$.*

PROOF. 1 follows from 2 and Proposition 18.27.

2. By Remark 18.21, we can assume $A = \emptyset$. Let $E = \overset{\text{Ls}}{\equiv}$ and consider $e = a_E$, a bounded hyperimaginary. We will show that we can eliminate e using Proposition 18.2. Let $p(x) = \text{tp}(a)$. We use point 3 of Proposition 18.26 to show that $E \upharpoonright p$ is an intersection of 0-definable equivalence relations. Let $\varphi(x, y) \in E \upharpoonright p$. We need to find $\varphi'(x, y) \in E \upharpoonright p$ such that $\varphi'(x, y) \vdash \varphi(x, y)$ and $E(x, x') \wedge E(y, y') \wedge \varphi'(x, y) \vdash \varphi'(x', y')$. Choose $\bar{\varphi} \in E(x, y) \upharpoonright$

p such that

$$\bar{\varphi}(x, y) \wedge \bar{\varphi}(y, z) \wedge \bar{\varphi}(z, u) \wedge \bar{\varphi}(u, v) \vdash \varphi(x, v).$$

Consider the following binary relation $R(b, c)$ on realizations b, c of p:

$$R(b, c) \Leftrightarrow \bar{\varphi}(x, b') \wedge \bar{\varphi}(x, c') \text{ does not fork over } \emptyset \text{ for some } b', c' \models p$$
$$\text{such that } E(b, b'), E(c, c'), \text{ and } b' \perp c'.$$

We will check that R is definable by some formula φ' as above. Since $e \in \operatorname{bdd}(\emptyset)$, $a \perp e$ and hence the type $E(x, a)$ does not fork over $\emptyset$. Likewise, for any $b \models p$ the type $E(x, b)$ does not fork over $\emptyset$. This implies that we can find an independent sequence b_1, b_2, b_3 in the E-class b_E, which shows that $\bar{\varphi}(x, b_1) \wedge \bar{\varphi}(x, b_2)$ does not fork over $\emptyset$. It follows that whenever $b, c \models p$ and $E(b, c)$ then $R(b, c)$. By choice of $\bar{\varphi}$, whenever $R(b, c)$ then $\models \varphi(b, c)$. Finally, it is obvious that if $E(b, b')$, $E(c, c')$, and $R(b, c)$, then $R(b', c')$.

To check the definability of R we show that R and its complement are type-definable. Type-definability of R follows from Lemma 18.28. For the complement $\bar{R}$ of R we need to use the lowness of T. First note that, since $E = \overset{\text{Ls}}{\equiv}$, using Corollary 10.8 it is easy to see that for all b, c, b', c' realizing p, if $b \perp c$, $b' \perp c'$, $E(b, b')$, and $E(c, c')$, if $\bar{\varphi}(x, b) \wedge \bar{\varphi}(x, c)$ does not fork over $\emptyset$, then also $\bar{\varphi}(x, b') \wedge \bar{\varphi}(x, c')$ does not fork over $\emptyset$. Hence for $b, c \models p$, $\bar{R}(b, c)$ if and only if there are $b', c' \models p$ such that $E(b, b')$, $E(c, c')$, $b' \perp c'$ and $\bar{\varphi}(x, b') \wedge \bar{\varphi}(x, c')$ forks over $\emptyset$. By Proposition 18.19 and Corollary 5.24 it is easily seen that this relation is type-definable over $\emptyset$. $\dashv$

Chapter 19

ORTHOGONALITY AND ANALYSABILITY

Definition 19.1. Let $p(x) \in S_E(e)$, $q(y) \in S_F(d)$ where e, d are hyperimaginaries. If $e, d \in \mathrm{dcl}^{\mathrm{eq}}(h)$, we say that *$p$, q are orthogonal over h* and we write $p \perp_h q$ if , if whenever $p'(x) \in S_E(h)$ is a nonforking extension of p and $q'(y) \in S_F(h)$ is a nonforking extension of q, then $a \mathop{\smile\!\!\!\!\!\!|}_h b$ for all $a \models p'$, for all $b \models q'$. We say that *p, q are orthogonal* and we write $p \perp q$ if $p \perp_h q$ for every hyperimaginary h over which e, d are definable.

Lemma 19.2. *Let T be simple, let e, d be hyperimaginaries and let $p(x) \in S_E(e)$, $q(y) \in S_F(d)$. If for every hyperimaginary h such that $e, d \in \mathrm{dcl}^{\mathrm{heq}}(h)$ there is some set C such that $h \in \mathrm{dcl}^{\mathrm{heq}}(C)$ and $p \perp_C q$, then $p \perp q$.*

Proof. Let $e, d \in \mathrm{dcl}^{\mathrm{heq}}(h)$ and let $h \in \mathrm{dcl}^{\mathrm{heq}}(C)$ be such that $p \perp_C q$. Let $p'(x), q'(x)$ be nonforking extensions of p, q over h and let $a \models p'$, $b \models q'$. Let $a'b' \equiv_h ab$ be such that $a'b' \mathop{\smile\!\!\!\!\!\!|}_h C$. Then a', b' realize nonforking extensions of p, q over C and, by assumption, $a' \mathop{\smile\!\!\!\!\!\!|}_C b'$. It follows that $a' \mathop{\smile\!\!\!\!\!\!|}_h b'$ and therefore $a \mathop{\smile\!\!\!\!\!\!|}_h b$. ⊣

Proposition 19.3. *Let T be simple, let e, d be hyperimaginaries and let $p(x) \in S_E(e)$, $q(y) \in S_F(d)$. Let e' and d' be hyperimaginaries such that $e \in \mathrm{dcl}^{\mathrm{heq}}(e')$ and $d \in \mathrm{dcl}^{\mathrm{heq}}(d')$. Then $p \perp q$ if and only if $p' \perp q'$ for all nonforking extensions $p' \in S_E(e')$ and $q' \in S_F(d')$ of p, q respectively.*

Proof. From left to right it is clear. Use Lemma 19.2 for the opposite direction. ⊣

Lemma 19.4. *Let T be simple and let e, d be hyperimaginaries. If $\mathrm{tp}(a/e)$ is orthogonal to $p(x) \in S_E(d)$ and $a' \in \mathrm{bdd}(ae)$, then $\mathrm{tp}(a'/e) \perp p$.*

Proof. Easy to check. ⊣

Proposition 19.5. *Let T be simple and let e, b be hyperimaginaries. Assume $a = (a_i : i \in I)$ is a sequence of hyperimaginaries and it is independent over e. If $\mathrm{tp}(a_i/e) \perp \mathrm{tp}(b/e)$ for all $i \in I$, then $\mathrm{tp}(a/e) \perp \mathrm{tp}(b/e)$.*

Proof. Assume that d is a hyperimaginary such that $e \in \mathrm{dcl}^{\mathrm{heq}}(d)$ and let $a' = (a'_i : i \in I) \equiv_e a$ and $a' \mathop{\smile\!\!\!\!\!\!|}_e d$ and let $b' \equiv_e b$ and $b' \mathop{\smile\!\!\!\!\!\!|}_e d$. Then a' is independent over d. By induction on $n < \omega$ it is easy to check that whenever $i_1, \ldots, i_n \in I$, $b' \mathop{\smile\!\!\!\!\!\!\mid}_d a'_{i_1} \ldots a'_{i_n}$, which implies $b' \mathop{\smile\!\!\!\!\!\!\mid}_d a'$. ⊣

LEMMA 19.6. *Let T be simple and let A be a set. Assume $(a_i : 1 \leq i \leq m)$ is A-independent, and each $\mathrm{tp}(a_i/A)$ is orthogonal to all types of* D*-rank* $< \alpha$. *If* $\mathrm{D}(\mathrm{tp}(d_j/A)) \leq \alpha$ *for all* $j = 1, \ldots, n$, *and* $d_1, \ldots, d_n \not\mathop{\smile\!\!\!\!\!\!\mid}_A a_i$ *for all* $i = 1, \ldots, m$, *then* $m \leq n$.

PROOF. The proof is by induction on n. We use Remark 14.6 several times. The starting case is $n = 1$. Assume $d_1 \not\mathop{\smile\!\!\!\!\!\!\mid}_A a_1$ and $d_1 \not\mathop{\smile\!\!\!\!\!\!\mid}_A a_2$. Then $\mathrm{D}(\mathrm{tp}(d_1/Aa_2)) < \alpha$ and therefore $\mathrm{tp}(a_1/A) \perp \mathrm{tp}(d_1/Aa_2)$. Since $a_1 \mathop{\smile\!\!\!\!\!\!\mid}_A a_2$, by orthogonality $a_1 \mathop{\smile\!\!\!\!\!\!\mid}_{Aa_2} d_1$ and therefore $a_1 \mathop{\smile\!\!\!\!\!\!\mid}_A d_1$, a contradiction.

Now consider the case $n + 1$. Assume $d_1, \ldots, d_{n+1} \not\mathop{\smile\!\!\!\!\!\!\mid}_A a_i$ for all $i \leq m$, and assume $m > n + 1$. We claim that $d_j \not\mathop{\smile\!\!\!\!\!\!\mid}_A a_1, \ldots, a_{n+1}$ for all $j \leq n + 1$. Assume $d_j \mathop{\smile\!\!\!\!\!\!\mid}_A a_1, \ldots, a_{n+1}$ for some $j \leq n + 1$, say for $j = n + 1$. Then $d_1, \ldots, d_n \not\mathop{\smile\!\!\!\!\!\!\mid}_{Ad_{n+1}} a_i$ for all $i \leq n + 1$. This contradicts the induction hypothesis, since $a_1, \ldots, a_{n+1}$ are independent over Ad_{n+1}, $\mathrm{tp}(a_i/Ad_{n+1})$ is orthogonal to every type of D-rank $< \alpha$ for all $i \leq n$, and $\mathrm{D}(\mathrm{tp}(d_j/Ad_{n+1})) \leq \alpha$ for all $j \leq n$. Now we claim that $a_{n+2} \mathop{\smile\!\!\!\!\!\!\mid}_A a_1, \ldots, a_{n+1} d_1, \ldots, d_j$ for all $j \leq n + 1$, a contradiction with the assumption for the case $j = n + 1$. We prove it by induction on j. Consider the case $j = 1$. Since $\mathrm{D}(\mathrm{tp}(d_1/Aa_1, \ldots, a_{n+1})) < \alpha$ and $a_{n+2} \mathop{\smile\!\!\!\!\!\!\mid}_A a_1, \ldots, a_{n+1}$, by orthogonality $a_{n+2} \mathop{\smile\!\!\!\!\!\!\mid}_{Aa_1, \ldots, a_{n+1}} d_1$. It follows that $a_{n+2} \mathop{\smile\!\!\!\!\!\!\mid}_A a_1, \ldots, a_{n+1} d_1$. Now consider the case $j + 1$, assuming $a_{n+2} \mathop{\smile\!\!\!\!\!\!\mid}_A a_1, \ldots, a_{n+1} d_1, \ldots, d_j$. Since $d_{j+1} \not\mathop{\smile\!\!\!\!\!\!\mid}_A a_1, \ldots, a_{n+1} d_1, \ldots, d_j$, it follows that $\mathrm{D}(d_{j+1}/Aa_1, \ldots, a_{n+1} d_1, \ldots, d_j)) < \alpha$. By orthogonality we obtain that $a_{n+2} \mathop{\smile\!\!\!\!\!\!\mid}_{Aa_1, \ldots, a_{n+1} d_1, \ldots, d_j} d_{j+1}$. By transitivity we conclude that $a_{n+2} \mathop{\smile\!\!\!\!\!\!\mid}_A a_1, \ldots, a_{n+1} d_1, \ldots, d_{j+1}$. ⊣

DEFINITION 19.7. Let $\boldsymbol{P}$ be a family (possibly a proper class) of partial types and let e be a hyperimaginary. Assume that $\boldsymbol{P}$ is e-invariant, that is, for every $f \in \mathrm{Aut}(\mathfrak{C}/e)$, for every $\pi \in \boldsymbol{P}$, $\pi^f \in \boldsymbol{P}$. A hyperimaginary type $p(x) \in S_E(e)$ is *internal in* $\boldsymbol{P}$ if for every hyperimaginary $a \models p$ there is some set B such that $e \in \mathrm{dcl}^{\mathrm{heq}}(B)$, and for some some sequence $(\pi_i : i \in I)$ of partial types $\pi_i \in \boldsymbol{P}$ over B there is a sequence $c = (c_i : i \in I)$ of realizations $c_i \models \pi_i$ such that

$$a \mathop{\smile\!\!\!\!\!\!\mid}_e B \text{ and } a \in \mathrm{dcl}^{\mathrm{heq}}(B, c).$$

The type $p(x) \in S_E(e)$ is *foreign to* $\boldsymbol{P}$ if for every hyperimaginary $a \models p$, for every set A such that $e \in \mathrm{dcl}^{\mathrm{heq}}(A)$ and $a \mathop{\smile\!\!\!\!\!\!\mid}_e A$, for every tuple c of realizations of types in $\boldsymbol{P}$ over A, $a \mathop{\smile\!\!\!\!\!\!\mid}_A c$.

Notice that in both cases it suffices to find some realization a of p with the corresponding properties.

LEMMA 19.8. *Let T be simple, and let c, d be hyperimaginaries. Let $p(x) \in S_E(e)$, let $e \in \mathrm{dcl}^{\mathrm{heq}}(d)$, and let $q(x) \in S_E(d)$ be a nonforking extension of p. Let $\boldsymbol{P}$ be an e-invariant family of partial types.*

1. *p is internal in $\boldsymbol{P}$ if and only if q is internal in $\boldsymbol{P}$.*
2. *If p is foreign to $\boldsymbol{P}$, then p' is foreign to $\boldsymbol{P}$.*
3. *If p is an amalgamation base and p' is foreign to $\boldsymbol{P}$, then p is foreign to $\boldsymbol{P}$.*

PROOF. Easy exercise. ⊣

REMARK 19.9. *Let T be simple let a, e be hyperimaginaries and let $\boldsymbol{P}$ be an e-invariant family of partial types. If $\mathrm{tp}(a/e)$ is internal in $\boldsymbol{P}$ and foreign to $\boldsymbol{P}$, then $a \in \mathrm{bdd}(e)$.*

PROOF. Since $\mathrm{tp}(a/e)$ is internal in $\boldsymbol{P}$, there is some set A such that $e \in \mathrm{dcl}^{\mathrm{heq}}(A)$ and $a \mathop{\smile\!\!\!\!\!\!|}_e A$, and there is some tuple c of realizations of types in $\boldsymbol{P}$ over A such that $a \in \mathrm{dcl}^{\mathrm{heq}}(A, c)$. Since $\mathrm{tp}(a/e)$ is foreign to $\boldsymbol{P}$, $a \mathop{\smile\!\!\!\!\!\!|}_A c$ and therefore $a \mathop{\smile\!\!\!\!\!\!|}_A a$. Then $a \mathop{\smile\!\!\!\!\!\!|}_e a$ and hence $a \in \mathrm{bdd}(e)$. ⊣

LEMMA 19.10. *Let T be simple, let a, b, e be hyperimaginaries, and let $\boldsymbol{P}$ be an e-invariant family of partial types.*

1. *If $\mathrm{tp}(a/e)$ is internal in $\boldsymbol{P}$ and $b \in \mathrm{dcl}^{\mathrm{heq}}(ae)$, then $\mathrm{tp}(b/e)$ is internal in $\boldsymbol{P}$.*
2. *If $\mathrm{tp}(a/e)$ is internal in $\boldsymbol{P}$, then $\mathrm{tp}(a/eb)$ is internal in $\boldsymbol{P}$.*

PROOF. Easy to check. ⊣

LEMMA 19.11. *Let T be simple, let a, b, e be hyperimaginaries and assume $\boldsymbol{P}$ is an e-invariant family of partial types and $\mathrm{tp}(a/e)$ is internal in $\boldsymbol{P}$. If $a' = \mathrm{Cb}(a/eb)$ then $\mathrm{tp}(a'/e)$ is internal in $\boldsymbol{P}$.*

PROOF. Choose a set A and a tuple c of realizations of types in $\boldsymbol{P}$ over A such that $e \in \mathrm{dcl}^{\mathrm{heq}}(A)$, $a \mathop{\smile\!\!\!\!\!\!|}_e A$ and $a \in \mathrm{dcl}^{\mathrm{heq}}(A, c)$. Let $(a_i : i < \omega)$ be a Morley sequence in $\mathrm{tp}(a/be)$. By Proposition 17.24 $a' \in \mathrm{dcl}^{\mathrm{heq}}(a_i : i < \omega)$. For each $i < \omega$ choose c_i, A_i such that $acA \equiv_e a_i c_i A_i$. Then c_i is a tuple of realizations of types in $\boldsymbol{P}$ over A_i, $e \in \mathrm{dcl}^{\mathrm{heq}}(A_i)$, $a_i \mathop{\smile\!\!\!\!\!\!|}_e A_i$ and $a_i \in \mathrm{dcl}^{\mathrm{heq}}(A_i, c_i)$. If necessary, we replace each A_i by $A'_i \equiv_{ea_i} A_i$ such that $A'_i \mathop{\smile\!\!\!\!\!\!|}_e A'_{<i}(a_i : i < \omega)$ (and then we choose c'_i such that $A'_i c'_i \equiv_{ea_i} c_i A_i$) to guarantee that $(A_i : i < \omega) \mathop{\smile\!\!\!\!\!\!|}_e (a_i : i < \omega)$. This implies $a' \mathop{\smile\!\!\!\!\!\!|}_e A$ where $A = \bigcup_{i<\omega} A_i$. Since $a' \in \mathrm{dcl}^{\mathrm{heq}}(A, (c_i : i < \omega))$, $\mathrm{tp}(a'/e)$ is internal in $\boldsymbol{P}$. ⊣

LEMMA 19.12. *Let T be simple, let a, e be hyperimaginaries and assume $\boldsymbol{P}$ is an e-invariant family of partial types. If there is some $b \in \mathrm{bdd}(ae) \setminus \mathrm{bdd}(e)$ such that $\mathrm{tp}(b/e)$ is internal in $\boldsymbol{P}$, then there is some $d \in \mathrm{dcl}^{\mathrm{heq}}(ae) \setminus \mathrm{bdd}(e)$ such that $\mathrm{tp}(d/e)$ is internal in $\boldsymbol{P}$.*

PROOF. Consider the orbit $X = \{f(b) : f \in \mathrm{Aut}(\mathfrak{C}/ae)\}$. By Proposition 15.28, there is a hyperimaginary d such that for every automorphism $f \in \mathrm{Aut}(\mathfrak{C})$, $f(d) = d$ if and only if $f(X) = X$. Fix an enumeration

$X = \{b_i : i < \alpha\}$ for some ordinal α. Each $\mathrm{tp}(b_i/e)$ is internal in $\boldsymbol{P}$ and hence we can find some set B_i such that $e \in \mathrm{dcl}^{\mathrm{heq}}(B_i)$ and $b_i \mathop{\smile\!\!\!\!\!\mid}_e B_i$ and there is some tuple c_i of realizations of types in $\boldsymbol{P}$ over B_i such that $b_i \in \mathrm{dcl}^{\mathrm{heq}}(B_i, c_i)$. Choose inductively $B'_i \equiv_{eb_i} B_i$ such that $B'_i \mathop{\smile\!\!\!\!\!\mid}_{eb_i} (b_j : j < \alpha) B'_{<i}$. Then $(b_i : i < \alpha) \mathop{\smile\!\!\!\!\!\mid}_e (B'_i : i < \alpha)$. Since $d \in \mathrm{dcl}^{\mathrm{heq}}(b_i : i < \alpha)$, if $B = \bigcup_{i<\alpha} B'_i$, then $d \mathop{\smile\!\!\!\!\!\mid}_e B$. Now choose c'_i such that $B'_i c'_i \equiv_{eb_i} B_i c_i$. It follows that each c'_i is a tuple of realizations of types in $\boldsymbol{P}$ over B'_i and $b_i \in \mathrm{dcl}^{\mathrm{heq}}(B'_i, c'_i)$. Hence d is definable over $B(c'_i : i < \alpha)$ and $\mathrm{tp}(d/e)$ is internal in $\boldsymbol{P}$. Clearly $d \in \mathrm{dcl}^{\mathrm{heq}}(ae)$. Since the orbit X is bounded, $b \in \mathrm{bdd}(d)$. Since $b \notin \mathrm{bdd}(e)$, $d \notin \mathrm{bdd}(e)$. ⊣

PROPOSITION 19.13. *Let T be simple. Let e, a be hyperimaginaries and let $\boldsymbol{P}$ be an e-invariant family of partial types. If $p(x) = \mathrm{tp}(a/e)$ is not foreign to $\boldsymbol{P}$, then there is some hyperimaginary $d \in \mathrm{dcl}^{\mathrm{heq}}(ae) \smallsetminus \mathrm{bdd}(e)$ such that $\mathrm{tp}(d/e)$ is internal in $\boldsymbol{P}$.*

PROOF. Assume $p(x) = \mathrm{tp}(a/e)$ is not foreign to $\boldsymbol{P}$, an e-invariant family of partial types. Then for some set A, $e \in \mathrm{dcl}^{\mathrm{heq}}(A)$, $a \mathop{\smile\!\!\!\!\!\mid}_e A$, and there is some tuple c of realizations of types in $\boldsymbol{P}$ over A such that $a \not\mathop{\smile\!\!\!\!\!\mid}_A c$. Let a_0 be a hyperimaginary equivalent to $\mathrm{Cb}(Ac/ae)$ (after fixing an enumeration of A). Then $a_0 \in \mathrm{bdd}(ae)$. Choose a Morley sequence $(A_i c_i : i < \omega)$ in $\mathrm{tp}(Ac/\mathrm{bdd}(ae))$. By Proposition 17.24, $a_0 \in \mathrm{dcl}^{\mathrm{heq}}((A_i c_i : i < \omega))$. Note that each c_i is a tuple of realizations of types in $\boldsymbol{P}$ over A_i. It is clear that $a \mathop{\smile\!\!\!\!\!\mid}_e A_i$ for all $i < \omega$. An induction shows that $a \mathop{\smile\!\!\!\!\!\mid}_e (A_i : i < n)$ for all $n < \omega$ and hence $a \mathop{\smile\!\!\!\!\!\mid}_e (A_i : i < \omega)$ and $a_0 \mathop{\smile\!\!\!\!\!\mid}_e (A_i : i < \omega)$. Hence $\mathrm{tp}(a_0/e)$ is internal in $\boldsymbol{P}$. If $a_0 \in \mathrm{bdd}(A)$ then (since by definition of canonical base $Ac \mathop{\smile\!\!\!\!\!\mid}_{a_0} ae$) $Ac \mathop{\smile\!\!\!\!\!\mid}_A ae$ and hence $c \mathop{\smile\!\!\!\!\!\mid}_A a$, contrarily to our assumption. Hence $a_0 \notin \mathrm{bdd}(A)$ and in particular $a_0 \notin \mathrm{bdd}(e)$. The existence of $d \in \mathrm{dcl}^{\mathrm{heq}}(ae) \smallsetminus \mathrm{bdd}(e)$ with $\mathrm{tp}(d/e)$ internal in $\boldsymbol{P}$ can now be justified by Lemma 19.12. ⊣

NOTATION 19.14. *For any ordinal α, let $\boldsymbol{P}_\alpha$ be the family of all formulas of D-rank α. Notice that $\boldsymbol{P}_\alpha$ is invariant and hence it is e-invariant for all hyperimaginaries e. We also put $\boldsymbol{P}_{\leq\alpha} = \bigcup_{i\leq\alpha} \boldsymbol{P}_i$.*

COROLLARY 19.15. *Let T be simple. Let e, a be hyperimaginaries. If α is the minimal ordinal such that $\mathrm{tp}(a/e) \not\perp p$ for some type p of D-rank α, then there is some hyperimaginary $d \in \mathrm{dcl}^{\mathrm{heq}}(ae) \smallsetminus \mathrm{bdd}(e)$ such that $\mathrm{tp}(d/e)$ is internal in the family $\boldsymbol{P}_\alpha$ of all formulas of D-rank α.*

PROOF. Since any extension of p has D-rank $\leq \alpha$, we may assume p is a type over a set of real parameters A such that $e \in \mathrm{dcl}^{\mathrm{heq}}(A)$. Let $\boldsymbol{P}$ be the family of all e-conjugates of p. Now, $\mathrm{tp}(a/e)$ is nonorthogonal to any type in $\boldsymbol{P}$ and therefore it is not foreign to $\boldsymbol{P}$. By Proposition 19.13, there is some

$d \in \operatorname{dcl}^{\text{heq}}(ae) \setminus \operatorname{bdd}(e)$ such that $\operatorname{tp}(d/e)$ is internal in $\boldsymbol{P}$. Hence $\operatorname{tp}(d/e)$ is also internal in $\boldsymbol{P}_\alpha$. ⊣

DEFINITION 19.16. Let e be a hyperimaginary and let $\boldsymbol{P}$ be an e-invariant family (as above, possibly a proper class) of partial types. A hyperimaginary type $p(x) \in S_E(e)$ is *analysable in* $\boldsymbol{P}$ if for every hyperimaginary $a \models p$ there is some ordinal α and some sequence $(a_i : i < \alpha) \in \operatorname{dcl}^{\text{heq}}(a, e)$ of hyperimaginaries a_i such that $\operatorname{tp}(a_i/ea_{<i})$ is internal in $\boldsymbol{P}$ for each $i < \alpha$ and $a \in \operatorname{bdd}(e(a_i : i < \alpha))$. If the analysis is in finitely many steps, say $(a_i : i \leq n)$, then, by Lemma 19.10, we can always assume $a_0 = e$, $a_i \in \operatorname{dcl}^{\text{heq}}(a_{i+1})$ and $a \in \operatorname{bdd}(a_n)$.

PROPOSITION 19.17. *Let T be supersimple. For all hyperimaginaries a, e such that* $\operatorname{SU}(a/e) < \infty$, *there is some ordinal α such that* $\operatorname{tp}(a/e)$ *is analyzable in finitely many steps in the family $\boldsymbol{P}_{\leq\alpha}$ of formulas of* D-*rank* $\leq \alpha$. *More precisely, for some $n < \omega$ there are hyperimaginaries $a_0, \dots, a_n \in \operatorname{dcl}^{\text{heq}}(a, e)$ and a sequence of ordinals $(\alpha_i : i < n)$ such that $e = a_0$, $a \in \operatorname{bdd}(a_n)$, and for each $i < n$, $a_i \in \operatorname{dcl}^{\text{heq}}(a_{i+1})$, $\operatorname{tp}(a_{i+1}/a_i)$ is internal in $\boldsymbol{P}_{\alpha_i}$ and $\operatorname{tp}(a/a_i)$ is orthogonal to every type of* D-*rank* $< \alpha_i$.

PROOF. The proof is by induction on $\alpha = \operatorname{SU}(a/e)$. If $\operatorname{SU}(a/e) = 0$, then $a \in \operatorname{bdd}(e)$ and the analysis only has the initial step $a_0 = e$. Assume now $\operatorname{SU}(a/e) = \alpha > 0$ and the result holds for all e' such that $\operatorname{SU}(a/e') < \alpha$. There is a finite tuple b such that $a \not\perp_e b$. Then $\operatorname{tp}(a/e) \not\perp \operatorname{tp}(b/e)$ and (see Proposition 19.3) $\operatorname{tp}(a/e) \not\perp \operatorname{tp}(b/A)$ where A is a set of real elements such that $e \in \operatorname{dcl}^{\text{heq}}(A)$ and $b \perp_e A$. Since T is supersimple and b is finite, $\operatorname{D}(\operatorname{tp}(b/A)) < \infty$. Let α_0 be the least ordinal such that $\operatorname{tp}(a/e)$ is nonorthogonal to a type $p(x)$ of D-rank α_0. By Corollary 19.15 there is some hyperimaginary $d \in \operatorname{dcl}^{\text{heq}}(ae) \setminus \operatorname{bdd}(e)$ such that $\operatorname{tp}(d/e)$ is internal in $\boldsymbol{P}_{\alpha_0}$. By minimality of α_0, $\operatorname{tp}(a/e)$ is orthogonal to all types of D-rank $< \alpha_0$. Let $a_1 = de$ and $a_0 = e$. By Lemma 19.10 $\operatorname{tp}(a_1/a_0)$ is internal in $\boldsymbol{P}_{\alpha_0}$. Since $d \in \operatorname{dcl}^{\text{heq}}(ae) \setminus \operatorname{bdd}(e)$, $a \not\perp_e d$ and so $a \not\perp_{a_0} a_1$ and $\operatorname{SU}(a/a_0) > \operatorname{SU}(a/a_1)$. By the induction hypothesis applied to $\operatorname{tp}(a/a_1)$ we obtain $a_2, \dots, a_n$ and $(\alpha_i : 0 < i < n)$ as required. ⊣

DEFINITION 19.18. Let T be supersimple. For all hyperimaginaries a, e we define the *analysability rank* of a over e as the least ordinal α (if there is one) for which there is an *analysis* in n steps $a_0, \dots, a_n$ of $\operatorname{tp}(a/e)$ as in Proposition 19.17 and the corresponding ordinals α_i are $\leq \alpha$. We denote it by $\operatorname{R}^{\text{an}}(a/e)$. If there is no such ordinal, we set $\operatorname{R}^{\text{an}}(a/e) = \infty$. By Proposition 19.17 $\operatorname{R}^{\text{an}}(a/e) < \infty$ if $\operatorname{SU}(a/e) < \infty$. Notice also that $\operatorname{R}^{\text{an}}(a/e) = 0$ if and only if $a \in \operatorname{bdd}(e)$.

LEMMA 19.19. *Let T be supersimple and let a, b, e be hyperimaginaries, and assume* $\operatorname{SU}(a/e) < \infty$.

1. $\mathrm{R}^{\mathrm{an}}(a/eb) \leq \mathrm{R}^{\mathrm{an}}(a/e)$.
2. *If* $a \mathop{\smile\!\!\!\!\!\mid}_e b$, *then* $\mathrm{R}^{\mathrm{an}}(a/eb) = \mathrm{R}^{\mathrm{an}}(a/e)$.

PROOF. 1. We proceed by induction on $\mathrm{SU}(a/eb)$ and we may assume $\mathrm{SU}(a/e) > 0$. We claim that for some $\alpha \leq \mathrm{R}^{\mathrm{an}}(a/e)$ there is some $d \in \mathrm{dcl}^{\mathrm{heq}}(aeb) \setminus \mathrm{bdd}(eb)$ such that $\mathrm{tp}(d/eb)$ is internal in $\boldsymbol{P}_\alpha$. Otherwise, whenever $\alpha \leq \mathrm{R}^{\mathrm{an}}(a/e)$, $d \in \mathrm{dcl}^{\mathrm{heq}}(aeb)$ and $\mathrm{tp}(d/eb)$ is internal in $\boldsymbol{P}_\alpha$, then $d \in \mathrm{bdd}(eb)$. Let $a_0, \dots, a_n$ with ordinals $(\alpha_i : i < n)$ be an analysis of $\mathrm{tp}(a/e)$ with $\alpha_i \leq \mathrm{R}^{\mathrm{an}}(a/e)$ for all $i < n$. Note that $a_0 = e \in \mathrm{bdd}(eb)$. Assume $a_i \in \mathrm{bdd}(eb)$. Since $a_{i+1} \in \mathrm{dcl}^{\mathrm{heq}}(aeb)$ and $\mathrm{tp}(a_{i+1}/a_i)$ is internal in $\boldsymbol{P}_{\alpha_i}$, by Lemma 19.10 $\mathrm{tp}(a_{i+1}/\mathrm{bdd}(eb))$ is internal in $\boldsymbol{P}_{\alpha_i}$ and by Lemma 19.8 $\mathrm{tp}(a_{i+1}/eb)$ is internal in $\boldsymbol{P}_{\alpha_i}$. By induction, $a_n \in \mathrm{bdd}(eb)$. Since $a \in \mathrm{bdd}(a_n)$, $a \in \mathrm{bdd}(eb)$ and therefore $\mathrm{R}^{\mathrm{an}}(a/be) = 0$. So, we may assume the claim is true. Choose β_0 minimal such that there is some $d \in \mathrm{dcl}^{\mathrm{heq}}(aeb) \setminus \mathrm{bdd}(eb)$ such that $\mathrm{tp}(d/eb)$ is internal in the family $\boldsymbol{P}_{\beta_0}$. By Corollary 19.15, $\mathrm{tp}(a/be)$ is orthogonal to every type of D-rank $< \beta_0$. Note that $a \not\mathop{\smile\!\!\!\!\!\mid}_{eb} d$ and hence $\mathrm{SU}(a/ebd) < \mathrm{SU}(a/eb)$. By the induction hypothesis $\mathrm{R}^{\mathrm{an}}(a/ebd) \leq \mathrm{R}^{\mathrm{an}}(a/e)$. A suitable analysis of $\mathrm{tp}(a/bde)$ after $b_0 = be$ and $b_1 = b_0 d$ with first ordinal β_0 gives an analysis of $\mathrm{tp}(a/be)$ with ordinals $\leq \mathrm{R}^{\mathrm{an}}(a/e)$.

2. The proof is by induction on $\mathrm{SU}(a/e)$. Let $\alpha = \mathrm{R}^{\mathrm{an}}(a/eb)$. We prove that $\mathrm{R}^{\mathrm{an}}(a/e) \leq \alpha$ assuming $a \mathop{\smile\!\!\!\!\!\mid}_e b$. Let $a_0, \dots, a_n$ be an analysis of $\mathrm{tp}(a/eb)$ with ordinals $(\alpha_i : i < n)$, all $\leq \alpha$. Assume for all $\beta \leq \alpha$, for all $d \in \mathrm{bdd}(ae)$ such that $\mathrm{tp}(d/e)$ is internal in $\boldsymbol{P}_\beta$, $d \in \mathrm{bdd}(e)$. We prove by induction on i that $a_i \mathop{\smile\!\!\!\!\!\mid}_e a$. This is clear for $a_0 = be$. Let $a'_{i+1} = \mathrm{Cb}(a_{i+1}/ae)$. Then $a'_{i+1} \in \mathrm{bdd}(ae)$. Since $\mathrm{tp}(a_{i+1}/a_i)$ is internal in $\boldsymbol{P}_{\alpha_i}$, by Lemma 19.11 $\mathrm{tp}(a'_{i+1}/a_i)$ is internal in $\boldsymbol{P}_{\alpha_i}$. By the induction hypothesis, $a_i \mathop{\smile\!\!\!\!\!\mid}_e a$. hence $a'_{i+1} \mathop{\smile\!\!\!\!\!\mid}_e a_i$ and by Lemma 19.8 $\mathrm{tp}(a'_{i+1}/e)$ is internal in $\boldsymbol{P}_{\alpha_i}$. By our assumption, $a'_{i+1} \in \mathrm{bdd}(e)$ and therefore $a_{i+1} \mathop{\smile\!\!\!\!\!\mid}_e a$. Since $a \in \mathrm{bdd}(a_n)$, we conclude that $a \mathop{\smile\!\!\!\!\!\mid}_e a$ and hence $a \in \mathrm{bdd}(e)$ and $\mathrm{R}^{\mathrm{an}}(a/e) = 0$. On the other hand, if the assumption does not hold, then by Lemma 19.12 for some ordinal $\beta \leq \alpha$ there is some $d \in \mathrm{dcl}(ae) \setminus \mathrm{bdd}(e)$ such that $\mathrm{tp}(d/e)$ is internal in $\boldsymbol{P}_\beta$. Choose β_0 minimal with this property and choose a corresponding d. By Corollary 19.15 $\mathrm{tp}(a/e)$ is orthogonal to every type of D-rank $< \beta_0$. Note that $a \not\mathop{\smile\!\!\!\!\!\mid}_e d$ and therefore $\mathrm{SU}(a/ed) < \mathrm{SU}(a/e)$. Since $a \mathop{\smile\!\!\!\!\!\mid}_{ed} b$, by the induction hypothesis $\mathrm{R}^{\mathrm{an}}(a/ed) \leq \mathrm{R}^{\mathrm{an}}(a/edb)$. By 1, $\mathrm{R}^{\mathrm{an}}(a/edb) \leq \alpha$. Adding to $b_0 = e$ and $b_1 = ed$ (and with first ordinal β_0) an analysis of $\mathrm{tp}(a/ed)$ with ordinals $\leq \alpha$ we obtain an analysis of $\mathrm{tp}(a/e)$ with ordinals $\leq \alpha$. ⊣

LEMMA 19.20. *Let T be supersimple and let a, b, e be hyperimaginaries. Assume* $\mathrm{SU}(a/e) < \infty$. *If* $b \in \mathrm{bdd}(ae)$, *then* $\mathrm{R}^{\mathrm{an}}(b/e) \leq \mathrm{R}^{\mathrm{an}}(a/e)$.

PROOF. By induction on $\mathrm{SU}(b/e)$. Let $\alpha = \mathrm{R}^{\mathrm{an}}(a/e)$ and let $a_0, \dots, a_n$ be an analysis of $\mathrm{tp}(a/e)$ with ordinals $(\alpha_i : i < n)$ such that $\alpha_i \leq \alpha$ for all

$i < n$. Since $a \in \mathrm{bdd}(a_n)$, if $b \mathop{\smile\!\!\!\!\!\!|}_e a_n$ then $b \mathop{\smile\!\!\!\!\!\!|}_e a$. In this case $b \subset \mathrm{bdd}(e)$ and therefore $\mathrm{R}^{\mathrm{an}}(b/e) = 0$. Consequently we may assume that there is a maximal $i < n$ such that $b \mathop{\smile\!\!\!\!\!\!|}_e a_i$. We claim that there is some ordinal $\beta \leq \alpha$ for which there is some $d \in \mathrm{bdd}(a_i b) \smallsetminus \mathrm{bdd}(a_i)$ such that $\mathrm{tp}(d/a_i)$ is internal in $\boldsymbol{P}_\beta$. Otherwise for all $\beta \leq \alpha$, if $d \in \mathrm{bdd}(a_i b)$ and $\mathrm{tp}(d/a_i)$ is internal in $\boldsymbol{P}_\beta$, then $d \in \mathrm{bdd}(a_i)$. We know that $\mathrm{tp}(a_{i+1}/a_i)$ is internal in the family $\boldsymbol{P}_{\alpha_i}$. Let $a'_{i+1} = \mathrm{Cb}(a_{i+1}/a_i b)$. By Lemma 19.11 $\mathrm{tp}(a'_{i+1}/a_i)$ is also internal in $\boldsymbol{P}_{\alpha_i}$. Since $a'_{i+1} \in \mathrm{bdd}(a_i b)$ and $\alpha_i \leq \alpha$, it is clear that $a'_{i+1} \in \mathrm{bdd}(a_i)$ and hence $a_{i+1} \mathop{\smile\!\!\!\!\!\!|}_{a_i} b$. Since $a_i \mathop{\smile\!\!\!\!\!\!|}_e b$, it follows that $a_{i+1} \mathop{\smile\!\!\!\!\!\!|}_e b$, in contradiction with the choice of a_i. By Lemma 19.12 for some $\beta \leq \alpha$ there is some $d \in \mathrm{dcl}^{\mathrm{heq}}(a_i b) \smallsetminus \mathrm{bdd}(a_i)$ such that $\mathrm{tp}(d/a_i)$ is internal in $\boldsymbol{P}_\beta$. Let $\beta_0 \leq \alpha$ be minimal with respect to the existence of some $d \in \mathrm{dcl}^{\mathrm{heq}}(a_i b) \smallsetminus \mathrm{bdd}(a_i)$ such that $\mathrm{tp}(d/a_i)$ is internal in $\boldsymbol{P}_{\beta_0}$. By Corollary 19.15 $\mathrm{tp}(b/a_i)$ is orthogonal to every type of D-rank $< \beta_0$. Since $b \not\mathop{\smile\!\!\!\!\!\!|}_{a_i} d$, $\mathrm{SU}(b/a_i d) < \mathrm{SU}(b/a_i)$ and, by the induction hypothesis, $\mathrm{R}^{\mathrm{an}}(b/a_i d) \leq \mathrm{R}^{\mathrm{an}}(a/a_i d)$. By Lemma 19.19 $\mathrm{R}^{\mathrm{an}}(a/a_i d) \leq \alpha$. Adding an analysis of $\mathrm{tp}(b/a_i d)$ witnessing $\mathrm{R}^{\mathrm{an}}(b/a_i d) \leq \alpha$ after the beginning $b_0 = a_i$ and $b_1 = a_i d$ (with first ordinal β_0) we obtain an analysis of $\mathrm{tp}(b/a_i)$ with ordinals $\leq \alpha$ showing $\mathrm{R}^{\mathrm{an}}(b/a_i) \leq \alpha$. Since $b \mathop{\smile\!\!\!\!\!\!|}_e a_i$, by Lemma 19.19 $\mathrm{R}^{\mathrm{an}}(b/e) = \mathrm{R}^{\mathrm{an}}(b/a_i) \leq \alpha$. ⊣

PROPOSITION 19.21. *Let T be supersimple and let a, e be hyperimaginaries. If $\mathrm{SU}(a/e) < \infty$, then $\mathrm{R}^{\mathrm{an}}(a/e)$ is the least ordinal α such that $\mathrm{tp}(a/e)$ is analysable in the family $\boldsymbol{P}_{\leq\alpha}$ in finitely many steps.*

PROOF. We assume that $a_0, \ldots, a_n \in \mathrm{dcl}^{\mathrm{heq}}(e, a)$, $a_0 = e$, $a \in \mathrm{bdd}(a_n)$, and for each $i < n$, $a_i \in \mathrm{dcl}^{\mathrm{heq}}(a_{i+1})$, $\mathrm{tp}(a_{i+1}/a_i)$ is internal in $\boldsymbol{P}_{\leq\alpha}$, and we prove that $\mathrm{R}^{\mathrm{an}}(a/e) \leq \alpha$. The proof is by induction on $\mathrm{SU}(a/e)$. We may clearly assume $\mathrm{SU}(a/e) > 0$. We may also assume $a_1 \notin \mathrm{bdd}(e)$ since otherwise, by lemmas 19.8 and 19.10, $\mathrm{tp}(a_2/e)$ is internal in $\boldsymbol{P}_{\leq\alpha}$ and hence we can delete a_1 in the analysis.

As in the proof of Proposition 19.17, $\mathrm{tp}(a/e)$ is nonorthogonal to some type of D-rank β for some ordinal β. Choose β minimal with this property. We claim that $\beta \leq \alpha$. Assume $\beta > \alpha$. By Lemma 19.4, $\mathrm{tp}(a_1/e)$ is orthogonal to every type of D-rank $< \beta$. Since $\mathrm{tp}(a_1/e)$ is internal in $\boldsymbol{P}_{\leq\alpha}$, there is some set A such that $e \in \mathrm{dcl}^{\mathrm{heq}}(A)$ and $a_1 \mathop{\smile\!\!\!\!\!\!|}_e A$ and some tuple c of realizations of formulas of D-rank $\leq \alpha$ over A such that $a_1 \in \mathrm{dcl}^{\mathrm{heq}}(A, c)$. We want to prove that $a_1 \mathop{\smile\!\!\!\!\!\!|}_A c$ and hence we may assume c is a finite tuple, say $c = c_1, \ldots, c_m$. Since $\mathrm{D}(\mathrm{tp}(c_j/Ac_{<j})) \leq \alpha < \beta$, by induction using orthogonality we get $a_1 \mathop{\smile\!\!\!\!\!\!|}_A c$. This implies $a_1 \mathop{\smile\!\!\!\!\!\!|}_A a_1$ and therefore $a_1 \mathop{\smile\!\!\!\!\!\!|}_e a_1$ and $a_1 \in \mathrm{bdd}(e)$, contrarily to our assumption. Consequently, $\beta \leq \alpha$.

By Corollary 19.15, there is some $d \in \mathrm{dcl}(ae) \smallsetminus \mathrm{bdd}(e)$ such that $\mathrm{tp}(d/e)$ is internal in $\boldsymbol{P}_\beta$. It follows that $a \not\mathop{\smile\!\!\!\!\!\!|}_e d$ and hence $\mathrm{SU}(a/de) < \mathrm{SU}(a/e)$.

Note that $a_0 d, a_1 d, \dots, a_n d$ witnesses that $\operatorname{tp}(a/ed)$ is analysable in $\boldsymbol{P}_{\le\alpha}$ in finitely many steps. By the induction hypothesis, $\mathrm{R}^{\mathrm{an}}(a/ed) \le \alpha$. Clearly, we can then start with $b_0 = e$ and $b_1 = ed$ (with first ordinal $\beta_0 = \beta$) and continue with a suitable analysis of $\operatorname{tp}(a/ed)$, say $b_1, \dots, b_k$ with ordinals $(\beta_i : 1 \le i < k)$ obtaining witnesses of $\mathrm{R}^{\mathrm{an}}(a/e) \le \alpha$. ⊣

LEMMA 19.22. *Let T be supersimple and let a, b, e be hyperimaginaries. If $\mathrm{SU}(ab/e) < \infty$, and $\mathrm{R}^{\mathrm{an}}(a/e), \mathrm{R}^{\mathrm{an}}(b/ae) \le \alpha$, then $\mathrm{R}^{\mathrm{an}}(ab/e) \le \alpha$.*

PROOF. Let $a_0, \dots, a_n$ be an analysis of $\operatorname{tp}(a/e)$ with ordinals $\le \alpha$ witnessing $\mathrm{R}^{\mathrm{an}}(a/e) \le \alpha$, and let $b_0, \dots, b_m$ be a corresponding analysis witnessing $\mathrm{R}^{\mathrm{an}}(b/Ae) \le \alpha$. Then $ae \in \operatorname{bdd}(a_n)$, $b_0 = ae$, and $ab \in \operatorname{bdd}(b_m)$. Take a set A such that $b_0 \in \operatorname{dcl}^{\mathrm{heq}}(A)$. Since $a_n \in \operatorname{dcl}^{\mathrm{heq}}(A)$ and $b_0 \mathop{\smile\!\!\!\!\!\!|}_{a_n} A$, it is clear that $\operatorname{tp}(b_0/a_n)$ is internal in any family $\boldsymbol{P}_\gamma$. Hence the sequence $a_0, \dots, a_n, b_0, \dots, b_m$ shows that $\operatorname{tp}(ab/e)$ is analysable in $\boldsymbol{P}_{\le\alpha}$. By Proposition 19.21, $\mathrm{R}^{\mathrm{an}}(ab/e) \le \alpha$. ⊣

LEMMA 19.23. *Let T be supersimple and let a, b, e be hyperimaginaries. If $\mathrm{SU}(ab/e) < \infty$, then $\mathrm{R}^{\mathrm{an}}(ab/e) = \max\{\mathrm{R}^{\mathrm{an}}(a/e), \mathrm{R}^{\mathrm{an}}(b/e)\}$.*

PROOF. By Lemma 19.20, $\mathrm{R}^{\mathrm{an}}(a/e), \mathrm{R}^{\mathrm{an}}(b/e) \le \mathrm{R}^{\mathrm{an}}(ab/e)$. By lemmas 19.19 and 19.22, if $\mathrm{R}^{\mathrm{an}}(a/e), \mathrm{R}^{\mathrm{an}}(b/e) \le \alpha$ also $\mathrm{R}^{\mathrm{an}}(ab/e) \le \alpha$. ⊣

PROPOSITION 19.24. *Let T be supersimple and let a, b, e be hyperimaginaries. If $\operatorname{tp}(b/e)$ is orthogonal to every type of D-rank $< \alpha$ and $\mathrm{R}^{\mathrm{an}}(a/e) < \alpha$, then $\operatorname{tp}(b/e) \perp \operatorname{tp}(a/e)$.*

PROOF. Let d be a hyperimaginary such that $e \in \operatorname{dcl}^{\mathrm{heq}}(d)$ and $b \mathop{\smile\!\!\!\!\!\!|}_e d$. We have to show that for every $a' \equiv_e a$ such that $a' \mathop{\smile\!\!\!\!\!\!|}_e d$ we have $b \mathop{\smile\!\!\!\!\!\!|}_d a'$. Note that, by Lemma 19.19, $\mathrm{R}^{\mathrm{an}}(a'/d) = \mathrm{R}^{\mathrm{an}}(a'/e) = \mathrm{R}^{\mathrm{an}}(a/e) < \alpha$. Hence, there is an analysis $a_0, \dots, a_n$ of $\operatorname{tp}(a'/d)$ with ordinals $\alpha_i < \alpha$. In particular, $a_0 = d$ and $a' \in \operatorname{bdd}(a_n)$. If $b \mathop{\smile\!\!\!\!\!\!|}_d a_n$ then $b \mathop{\smile\!\!\!\!\!\!|}_d a'$ and we have finished. Otherwise there is a maximal i such that $b \mathop{\smile\!\!\!\!\!\!|}_d a_i$. Then $b \mathop{\smile\!\!\!\!\!\!|}_e a_i$. Since $\operatorname{tp}(a_{i+1}/a_i)$ is internal in the family $\boldsymbol{P}_{\alpha_i}$, there is a set A such that $a_i \in \operatorname{dcl}^{\mathrm{heq}}(A)$ and $a_{i+1} \mathop{\smile\!\!\!\!\!\!|}_{a_i} A$ and a tuple c of realizations of formulas over A of D-rank α_i such that $a_{i+1} \in \operatorname{dcl}^{\mathrm{heq}}(A, c)$. We may assume $A \mathop{\smile\!\!\!\!\!\!|}_{a_{i+1}} b$. Then $A \mathop{\smile\!\!\!\!\!\!|}_{a_i} b a_{i+1}$ and thus $b \mathop{\smile\!\!\!\!\!\!|}_e A$. We want to check that $b \mathop{\smile\!\!\!\!\!\!|}_A c$, and to do so we may assume c is a finite tuple, say $c = c_1, \dots, c_m$, where for every $j = 1, \dots, m$, $\mathrm{D}(\operatorname{tp}(c_j/A)) < \alpha$. By our orthogonality assumption one easily sees by induction that $b \mathop{\smile\!\!\!\!\!\!|}_A c_1, \dots, c_j$ for every $j \le m$. In particular, $b \mathop{\smile\!\!\!\!\!\!|}_A c$. Consequently, $b \mathop{\smile\!\!\!\!\!\!|}_A a_{i+1}$ and $b \mathop{\smile\!\!\!\!\!\!|}_d a_{i+1}$, a contradiction. ⊣

PROPOSITION 19.25. *Let T be supersimple, let e be a hyperimaginary, and let $a = (a_i : i \in I)$ be a sequence of hyperimaginaries a_i. If $\mathrm{SU}(a/e) < \infty$ and $\mathrm{R}^{\mathrm{an}}(a_i/e) < \alpha$ for each $i \in I$, then $\mathrm{R}^{\mathrm{an}}(a/e) < \alpha$.*

PROOF. Since $\mathrm{SU}(a/e) < \infty$, by Proposition 16.27, there is a finite subset $I_0 \subseteq I$ such that $a \in \mathrm{bdd}(a \restriction I_0, e)$. By Lemma 19.20 it is then enough to prove the result for I finite, and this can be easily done by induction on $|I|$ using Lemma 19.23. $\dashv$

PROPOSITION 19.26. *Let T be supersimple, let a, e be hyperimaginaries and let a' be a hyperimaginary equivalent to some enumeration of*

$$D = \{b \in \mathrm{dcl}^{\mathrm{heq}}(a, e) : \mathrm{R}^{\mathrm{an}}(b/e) < \alpha\}.$$

If $\mathrm{SU}(a/e) < \infty$, *then* $\mathrm{R}^{\mathrm{an}}(a'/e) < \alpha$ *and* $\mathrm{tp}(a/a')$ *is orthogonal to every type of* D-*rank* $< \alpha$.

PROOF. Notice that $\mathrm{SU}(a'/e) < \infty$. By Proposition 19.25, $\mathrm{R}^{\mathrm{an}}(a'/e) < \alpha$. Assume now $\mathrm{tp}(a/a')$ is nonorthogonal to some type of D-rank $\beta < \alpha$. By Corollary 19.15 there is some $d \in \mathrm{dcl}^{\mathrm{heq}}(aa') \smallsetminus \mathrm{bdd}(a')$ such that $\mathrm{tp}(d/a')$ is internal in $\boldsymbol{P}_\beta$. Let $a_0, \dots, a_n$ be an analysis of $\mathrm{tp}(a'/e)$ with ordinals $< \alpha$. Since $a' \in \mathrm{bdd}(a_n)$, by lemmas 19.8 and 19.10 $\mathrm{tp}(a_n d/a_n)$ is internal in $\boldsymbol{P}_\beta$. The tuples a_n and $a_n d$ provide an analysis of $\mathrm{tp}(d/a_n)$ in $\boldsymbol{P}_\beta$. By Proposition 19.21, $\mathrm{R}^{\mathrm{an}}(d/a_n) < \alpha$. Since also $\mathrm{R}^{\mathrm{an}}(a_n/e) < \alpha$, by Lemma 19.22, $\mathrm{R}^{\mathrm{an}}(da_n/e) < \alpha$. Since $a' \in \mathrm{dcl}(ae)$, we have $da_n \in \mathrm{dcl}(ae)$. Hence $da_n \in D$ and $da_n \in \mathrm{dcl}(a')$, in contradiction with $d \notin \mathrm{bdd}(a')$. $\dashv$

Chapter 20

HYPERIMAGINARIES IN SUPERSIMPLE THEORIES

LEMMA 20.1. *Let T be supersimple, let $E = E(x, y)$ be a 0-type-definable equivalence relation and let $q(x) \in S(\emptyset)$. There is a formula $\psi(x, y) \in E(x, y)$ such that for all $a, b \models q$, if $\models \psi(a, b)$ then* $\operatorname{bdd}(a_E) = \operatorname{bdd}(b_E)$.

PROOF. As usual, we assume that all formulas in $E(x, y)$ are symmetric. Let $\varphi(x, y) \in E(x, y)$ be a formula such that $\varphi(x, a)$ has smallest D-rank among all formulas in $E(x, y)$ for $a \models q$. Choose $\psi(x, y) \in E(x, y)$ such that $\psi^3(x, y) \vdash \varphi(x, y)$, that is,

$$\psi(x, z) \wedge \psi(z, u) \wedge \psi(u, y) \vdash \varphi(x, y).$$

Assume $a, b \models q$, $\models \psi(a, b)$ and $b_E \notin \operatorname{bdd}(a_E)$. By Proposition 16.4, there is an a_E-indiscernible sequence $([b_i]_E : i < \omega)$ of distinct hyperimaginaries $[b_i]_E$ starting with $[b_0]_E = b_E$. We can extend the sequence and apply the Erdős–Rado Theorem to obtain a formula $\chi(x, y) \in E$ such that $\models \neg\chi(b_i, b_j)$ for all $i < j < \omega$. We can assume $b = b_0$ and $b \equiv_{a_E} b_i$ for all $i < \omega$. Choose $\theta(x, y) \in E(x, y)$ such that

$$\theta(x, y) \vdash \psi(x, y) \text{ and } \theta^2(x, y) \vdash \chi(x, y).$$

For each $i < \omega$ there is some a_i such that $E(a_i, a)$ and $ab \equiv a_i b_i$. Hence $\models \psi(b_i, a_i)$ and therefore $\theta(x, b_i) \vdash \varphi(x, a)$. Moreover $\theta(x, b_i) \wedge \theta(x, b_j)$ is inconsistent for all $i < j < \omega$. We can extend the sequence $(b_i : i < \omega)$ and then apply Proposition 1.6, so we can assume $(b_i : i < \omega)$ is indiscernible over a. Then $\theta(x, b_0)$ divides over a, and by Remark 14.2, $\mathrm{D}(\theta(x, b_0)) < \mathrm{D}(\varphi(x, a))$. Since $b_0 \models q$, this contradicts the minimality in the choice of $\varphi(x, y)$. ⊣

LEMMA 20.2. *Assume T is supersimple, and α is an ordinal number such that for every hyperimaginary a_0, if there exists a sequence of imaginaries a with $\mathrm{SU}(a) < \infty$ and such that $a_0 \in \operatorname{dcl}^{\text{heq}}(a)$, and $\mathrm{R}^{\text{an}}(a/a_0) < \alpha$, then a_0 is eliminable. If a_0 is a hyperimaginary and there is a sequence of imaginaries a such that $\mathrm{SU}(a) < \infty$, and $a_0 \in \operatorname{dcl}^{\text{heq}}(a)$, and $\operatorname{tp}(a/a_0)$ is internal in the family $\boldsymbol{P}_\alpha$ of all formulas of D-rank α and it is orthogonal to every type of D-rank $< \alpha$, then a_0 is eliminable.*

PROOF. We will work in T^{eq}, so $\mathrm{acl} = \mathrm{acl}^{\mathrm{eq}}$. We start by changing a_0 by a more convenient hyperimaginary a_E.

By propositions 15.6, and 17.8, there is a 0-type-definable equivalence relation E such that $\mathrm{tp}(a/a_E)$ is an amalgamation base, $a_E \in \mathrm{bdd}(a_0)$, and $a_0 \in \mathrm{dcl}^{\mathrm{heq}}(a_E)$. Note that $a \mathop{\smile\!\!\!\!\!\mid}_{a_0} a_E$. By Proposition 19.3 $\mathrm{tp}(a/a_E)$ is orthogonal to every type of D-rank $< \alpha$ and by Lemma 19.8 $\mathrm{tp}(a/a_E)$ is internal in $\boldsymbol{P}_\alpha$. By Lemma 18.6 it suffices to show that a_E is eliminable.

We will find a sequence of imaginaries $\tilde{a}$ such that $\mathrm{SU}(\tilde{a}) < \infty$, $a_E \in \mathrm{dcl}^{\mathrm{heq}}(\tilde{a})$, and $\mathrm{R}^{\mathrm{an}}(\tilde{a}/a_E) < \alpha$. The assumption of the lemma will imply then that a_E is eliminable and the proof will be completed.

Let $q(x) = \mathrm{tp}(a)$. By Remark 15.17 $\mathrm{tp}(a/a_E) \equiv q(x) \wedge E(x, a)$. Since $\mathrm{SU}(a) < \infty$, by Proposition 16.27 there is a finite subtuple a^- of a such that $a \subseteq \mathrm{acl}(a^-)$. If $a = (a_i : i \in I)$, there is a finite subset $I^- \subseteq I$ such that $a^- = (a_i : i \in I^-)$. For any other sequence $b = (b_i : i \in I)$ we will understand that $b^- = (b_i : i \in I^-)$. In particular, if x is a tuple of variables corresponding to a, then x^- is the finite subtuple of x that corresponds to a^-. We can choose a partial type Σ over $\emptyset$ which defines $E \restriction q$, the restriction of E to realizations of q, and has the following additional properties:

1. If $\psi(x, y) \in \Sigma$, then $\psi(x, y) \vdash x_i \in \mathrm{acl}(x^-)$ and $\psi(x, y) \vdash y_i \in \mathrm{acl}(y^-)$ for all variables x_i, y_i appearing in ψ.
2. All formulas $\psi(x, y) \in \Sigma$ are symmetric.
3. If $\psi_0(x, y) \in E$ is as in Lemma 20.1 for E and q, then $\psi(x, y) \wedge \psi(y, z) \vdash \psi_0(x, z)$ for all $\psi(x, y) \in \Sigma(x, y)$.

By Lemma 19.10 $\mathrm{tp}(a^-/a_E)$ is internal in $\boldsymbol{P}_\alpha$ and by Lemma 19.4 it is orthogonal to all types of D-rank $< \alpha$. By internality, there is some set A such that $a_E \in \mathrm{dcl}^{\mathrm{heq}}(A)$ and for some $n < \omega$ there are formulas $\varphi_1(z_1), \ldots, \varphi_n(z_n)$ over A of D-rank α and realizations $c_1 \models \varphi_1, \ldots, c_n \models \varphi_n$ such that $a^- \mathop{\smile\!\!\!\!\!\mid}_{a_E} A$ and $a^- \in \mathrm{dcl}(A, c_1, \ldots, c_n)$. There is some formula $\chi = \chi(x^-, z_1, \ldots, z_n) \in L(A)$ such that $\models \forall z_1 \ldots z_n \exists^{=1} x^- \chi(x^-, z_1, \ldots, z_n)$ and $\models \chi(a^-, c_1, \ldots, c_n)$. We define

$$\chi_0(x^-) = \exists z_1 \ldots z_n (\chi(x^-, z_1 \ldots, z_n) \wedge \varphi_1(z_1) \wedge \cdots \wedge \varphi_n(z_n)).$$

Let $\tilde{a}$ be the tuple a extended by finitely many parameters from A such that each $\varphi_i(x^-)$ is over $\tilde{a}$ and χ is also over $\tilde{a}$. Again, for each tuple $b = (b_i : i \in I)$ of the length of a, by $\bar{b}$ we understand a corresponding tuple of the length of $\tilde{a}$, a finite extension of b. Let $\bar{q}(\bar{x}) = \mathrm{tp}(\tilde{a})$ and let us write $\chi_0(x^-) = \chi_0(x^-, \tilde{a})$, and $\varphi_i(z_i) = \varphi_i(z_i, \tilde{a})$ with $\chi_0(x^-, \bar{y}), \varphi_i(z_i, \bar{y}) \in L$.

We define D as the following class of hyperimaginaries:

$$D = \{d \in \mathrm{dcl}^{\mathrm{heq}}(\tilde{a}) : \mathrm{R}^{\mathrm{an}}(d/a_E) < \alpha\}.$$

There is some hyperimaginary interdefinable with D and by Proposition 15.6 there is some 0-type-definable equivalence relation F such that $\tilde{a}_F$ is equivalent to some sequence enumerating D. By Proposition 19.26 $\mathrm{R}^{\mathrm{an}}(\bar{a}_F/a_E) < \alpha$. By Lemma 19.20, $\mathrm{R}^{\mathrm{an}}(d/a_E) < \alpha$ if $d \in \mathrm{bdd}(\bar{a}_F)$. It follows that

$$D = \mathrm{dcl}^{\mathrm{heq}}(\bar{a}) \cap \mathrm{bdd}(\bar{a}_F).$$

By Proposition 17.8, $\mathrm{tp}(\bar{a}/\bar{a}_F)$ is an amalgamation base. Since $\bar{a}$ is a finite extension of a, it is contained in the algebraic closure of a finite tuple and by Remark 16.26 and supersimplicity, $\mathrm{SU}(\bar{a}) < \infty$. Since $a_E \in D$ (we may assume $\alpha > 0$) it is clear that $a_E \in \mathrm{dcl}^{\mathrm{heq}}(\bar{a}_F)$. By Proposition 19.26 $\mathrm{tp}(\bar{a}/\bar{a}_F)$ is orthogonal to every type of D-rank $< \alpha$. We recapitulate:

1. $\mathrm{tp}(\bar{a}/\bar{a}_F)$ is an amalgamation base.
2. $\mathrm{SU}(\bar{a}) < \infty$.
3. $a_E \in \mathrm{dcl}^{\mathrm{heq}}(\bar{a}_F)$.
4. $\mathrm{R}^{\mathrm{an}}(\bar{a}_F/a_E) < \alpha$.
5. $\mathrm{tp}(\bar{a}/\bar{a}_F)$ is orthogonal to every type of D-rank $< \alpha$.

All these properties still hold for any realization $\bar{a} \models \bar{q}$ and therefore we can forget about the particular realization of $\bar{q}$ we were discussing.

We define a relation $S_\psi = S_\psi(\bar{x}, y)$ for every $\psi(x, y) \in \Sigma$ as follows:

$S_\psi(\bar{a}, b)$ if and only if $\bar{a} \models \bar{q}$, and $b \models q$, and there is some $\bar{c} \models \bar{q}$ such that

- $F(\bar{c}, \bar{a})$ and $\bar{c} \mathop{\smile\!\!\!\!\!\!\;|}_{\bar{a}_F} b$,
- $\models \psi_0(c, b)$ and $\chi_0(x^-, \bar{c}) \wedge \psi(x, c) \wedge \psi(x, b)$ does not fork over $\bar{a}_F$.

We want to prove that S_ψ is definable on $\bar{q} \times q$, that is, there is a formula $\rho_\psi(\bar{x}, y) \in L$ such that for $\bar{a} \models \bar{q}$, for $b \models q$, $\models \rho_\psi(\bar{a}, b)$ if and only if $S_\psi(\bar{a}, b)$. For this it is enough to prove that S_ψ and its complement $\overline{S_\psi}$ are type-definable on $\bar{q} \times q$.

CLAIM 1. *S_ψ is type-definable on $\bar{q} \times q$.*

PROOF OF CLAIM 1. Let $\bar{a} \models \bar{q}$ and $b \models q$. By definition, $S_\psi(\bar{a}, b)$ if and only if for some $\bar{c} \models \bar{q}$, $F(\bar{c}, \bar{a})$, $\bar{c} \mathop{\smile\!\!\!\!\!\!\;|}_{\bar{a}_F} b$, $\models \psi_0(c, b)$ and $\chi_0(x^-, \bar{c}) \wedge \psi(x, c) \wedge \psi(x, b)$ does not fork over $\bar{a}_F$. Since $\bar{a}_F = \bar{c}_F$ and (by Remark 15.17) $\mathrm{tp}(\bar{a}/\bar{a}_F) \equiv \bar{q}(\bar{x}) \wedge F(\bar{x}, \bar{a})$ is constant for every $\bar{a} \models \bar{q}$, the condition $\bar{c} \mathop{\smile\!\!\!\!\!\!\;|}_{\bar{a}_F} b$ is type-definable by Corollary 16.41. We must check that

$$\chi_0(x^-, \bar{c}) \wedge \psi(x, c) \wedge \psi(x, b) \text{ does not fork over } \bar{a}_F$$

is also type-definable as a condition on $\bar{c}, b$ (recall $\bar{a}_F = \bar{c}_F$). By Proposition 16.23 it is equivalent to the existence of a Morley sequence $(\bar{c}_i, b_i : i < \omega)$ in $\mathrm{tp}(\bar{c}, b/\bar{c}_F)$ such that $\{\chi_0(x^-, \bar{c}_i) \wedge \psi(x, c_i) \wedge \psi(x, b_i) : i < \omega\}$ is consistent. A Morley sequence in $\mathrm{tp}(\bar{c}, b/\bar{c}_F)$ is a $\bar{c}_F$-indiscernible, $\bar{c}_F$-independent sequence of realizations of $\mathrm{tp}(\bar{c}, b/\bar{c}_F)$. Since $\mathrm{tp}(\bar{c}, b/\bar{c}_F)$ is not fixed for all $\bar{c} \models \bar{q}$ and $b \models q$, we cannot use Corollary 16.42. Indiscernibility is not a problem. Notice that $(\bar{c}_i b_i : i < \omega)$ is independent over $\bar{c}_F$ if and only if $\bar{c}_i \mathop{\smile\!\!\!\!\!\!\;|}_{\bar{c}_F} \bar{c}_{<i}(b_i : i < \omega)$ and $b_i \mathop{\smile\!\!\!\!\!\!\;|}_{\bar{c}_F} b_{<i}$ for all $i < \omega$. Now using Corollary 16.42 we can express $\bar{c}_i \mathop{\smile\!\!\!\!\!\!\;|}_{\bar{c}_F} \bar{c}_{<i}(b_i : i < \omega)$. We need only to show that assuming $\bar{a}_F = \bar{c}_F$ and $\models \psi_0(a,b)$ we can express

$$b_i \mathop{\smile\!\!\!\!\!\!\;|}_{\bar{a}_F} b_{<i}$$

by a partial type. Notice that, on $\bar{q}$, F implies E and then $a_E = c_E$. Since $\models \psi_0(a,b)$, the equivalence classes a_E and b_E are interbounded. By Lemma 19.19 $\mathrm{R}^{\mathrm{an}}(\bar{a}_F/b_E) < \alpha$ and therefore, by Proposition 19.24,

$$\mathrm{tp}(b/b_E) \perp \mathrm{tp}(\bar{a}_F/b_E).$$

By Proposition 19.5, $\mathrm{tp}((b_i : i < \omega)/b_E) \perp \mathrm{tp}(\bar{a}_F/b_E)$. Hence $b_i b_{<i} \mathop{\smile\!\!\!\!\!\!\;|}_{b_E} \bar{a}_F$ and by change of base (see Proposition 16.16) $b_i \mathop{\smile\!\!\!\!\!\!\;|}_{b_E} b_{<i}$ if and only if $b_i \mathop{\smile\!\!\!\!\!\!\;|}_{\bar{a}_F} b_{<i}$. Again by Corollary 16.41 we can express $b_i \mathop{\smile\!\!\!\!\!\!\;|}_{b_E} b_{<i}$ by a type. ⊣

CLAIM 2. *$\overline{S_\psi}$ is type-definable on $\bar{q} \times q$.*

PROOF OF CLAIM 2. Let $\bar{a} \models \bar{q}$ and $b \models q$. Notice first that $\overline{S_\psi}(\bar{a},b)$ if and only if there exists some $\bar{c} \models \bar{q}$ such that $F(\bar{a},\bar{c})$, $\bar{c} \mathop{\smile\!\!\!\!\!\!\;|}_{\bar{a}_F} b$ and either $\not\models \psi_0(c,b)$ or $\chi_0(x^-,\bar{c}) \wedge \psi(x,c) \wedge \psi(x,b)$ forks over $\bar{a}_F$. We check this. Assume first $S_\psi(\bar{a},b)$ and that there is some $\bar{c}$ as described. By definition of S_ψ there is another $\bar{d} \models \bar{q}$ such that $F(\bar{d},\bar{a})$, $\bar{d} \mathop{\smile\!\!\!\!\!\!\;|}_{\bar{a}_F} b$, $\models \psi_0(d,b)$, and $\chi_0(x^-,\bar{d}) \wedge \psi(x,d) \wedge \psi(x,b)$ does not fork over $\bar{a}_F$. Since $\bar{c} \stackrel{\mathrm{Ls}}{\equiv}_{\bar{a}_F} \bar{d}$, by Corollary 16.38 $\chi_0(x^-,\bar{c}) \wedge \psi(x,c) \wedge \psi(x,b)$ does not fork over $\bar{a}_F$ and hence $\not\models \psi_0(c,b)$. But $\psi(x,c) \wedge \psi(x,b)$ is consistent and by assumption $\psi(x,y)$ is symmetric and $\psi(x,y) \wedge \psi(y,z) \vdash \psi_0(x,z)$, which implies $\models \psi_0(c,b)$, a contradiction. The other direction is clear.

Condition $\bar{c} \mathop{\smile\!\!\!\!\!\!\;|}_{\bar{a}_F} b$ (assuming $\bar{a}_F = \bar{c}_F$) is easily seen to be type-definable by Corollary 16.41. A more difficult task is to show that condition

$$\chi_0(x^-,\bar{c}) \wedge \psi(x,c) \wedge \psi(x,b) \text{ forks over } \bar{a}_F$$

is type-definable. We can solve the problem proving that in fact, assuming $\models \psi_0(b,c)$, $F(\bar{a},\bar{c})$, and $\bar{c} \mathop{\smile\!\!\!\!\!\!\;|}_{\bar{a}_F} b$, the formula $\chi_0(x^-,\bar{c}) \wedge \psi(x,c) \wedge \psi(x,b)$ forks over $\bar{a}_F$ if and only if it $n+2$-divides over $\bar{a}_F$ (recall that n is the number of formulas witnessing internality). Clearly $n+2$-dividing over $\bar{a}_F$ is type-definable over any representative of $\bar{a}_F$.

We assume $\models \psi_0(b,c)$, $F(\bar{a},\bar{c})$, $\bar{c} \mathop{\smile\!\!\!\!\!\!\;|}_{\bar{a}_F} b$, and $\chi_0(x^-,\bar{c}) \wedge \psi(x,c) \wedge \psi(x,b)$ forks over $\bar{a}_F$ and we will prove that this formula $n+2$-divides over $\bar{a}_F$. This will finish the proof of the claim. Let $(\bar{c}_i b_i : i < \omega)$ be a Morley sequence in $\mathrm{tp}(\bar{c}b/\bar{a}_F)$. It will suffice to check inconsistency of

$$\{\chi_0(x^-,\bar{c}_i) \wedge \psi(x,c_i) \wedge \psi(x,b_i) : i < n+2\}.$$

Suppose it is consistent, and let e realize it. Let $0 < i < n+2$. Then $e \not\mathop{\smile\!\!\!\!\!\!\;|}_{\bar{a}_F} \bar{c}_i b_i$. Since $\bar{c}_i b_i \mathop{\smile\!\!\!\!\!\!\;|}_{\bar{a}_F} \bar{c}_0$ and $\bar{a}_F \in \mathrm{dcl}^{\mathrm{heq}}(\bar{c}_i)$, we get $e \not\mathop{\smile\!\!\!\!\!\!\;|}_{\bar{c}_0} \bar{c}_i b_i$. The formulas $\varphi_1(z_1,\bar{c}_1),\ldots,\varphi_n(z_n,\bar{c}_n)$ have D-rank α and, since $\models \chi(e^-,\bar{c}_0)$, there are $d_j \models \varphi_j(z_j, \bar{c}_0)$ such that $e \in \mathrm{dcl}(\bar{c}_0, d_1, \ldots, d_n)$. Since $\psi(x, y) \in \Sigma$ and $\models \psi(e, c_i)$, we know that $e \subseteq \mathrm{acl}(e^-)$. Hence $d_1, \ldots, d_n \not\perp_{\bar{c}_0} \bar{c}_i b_i$.

Now we claim that $\mathrm{tp}(\bar{c}_i b_i/\bar{c}_0)$ is orthogonal to all types of D-rank $< \alpha$ if $i > 0$. Note that $\bar{c}_i \perp_{\bar{c}_0} b_i$ (because $\bar{a}_F \in \mathrm{dcl}^{\mathrm{heq}}(\bar{c}_0)$, and $\bar{c}_i \perp_{\bar{a}_F} b_i$, and $\bar{c}_i b_i \perp_{\bar{a}_F} \bar{c}_0$) and then, by Proposition 19.3, it is enough to prove that $\mathrm{tp}(\bar{c}_i/\bar{c}_0)$ and $\mathrm{tp}(b_i/\bar{c}_0)$ are orthogonal to all types of D-rank $< \alpha$. Orthogonality of $\mathrm{tp}(\bar{c}_i/\bar{c}_0)$ is clear by Proposition 19.3 since $\bar{a}_F \in \mathrm{dcl}^{\mathrm{heq}}(\bar{c}_0)$, and $\bar{c}_i \perp_{\bar{a}_F} \bar{c}_0$, and $\mathrm{tp}(\bar{c}_i/\bar{a}_F)$ is orthogonal to all types of D-rank $< \alpha$. With respect to $\mathrm{tp}(b_i/\bar{c}_0)$, notice that since $\mathrm{tp}(b_i/b_E)$ is orthogonal to all types of D-rank $< \alpha$ and $\mathrm{R}^{\mathrm{an}}(\bar{a}_F/b_E) < \alpha$, by Proposition 19.22 $\mathrm{tp}(b_i/b_E) \perp \mathrm{tp}(\bar{a}_F/b_E)$ and therefore $b_i \perp_{b_E} \bar{a}_F$. Then $\mathrm{tp}(b_i/\bar{c}_0)$ is also orthogonal to all types of D-rank $< \alpha$, because $b_i \perp_{\bar{a}_F} \bar{c}_0$ and thus $b_i \perp_{b_E} \bar{c}_0$.

Lemma 19.6 implies that $d_1, \ldots, d_n \not\perp_{\bar{c}_0} \bar{c}_i b_i$ for at most n tuples $\bar{c}_i b_i$, a contradiction. $\dashv$

By the claims, there is a formula $\rho_\psi(\bar{x}, y) \in L$ such that for all $\bar{a} \models \bar{q}$, $b \models q$,

$$\models \rho_\psi(\bar{a}, b) \text{ if and only if } S_\psi(\bar{a}, b).$$

Note that for all $b \models q$ and all $\bar{a}, \bar{c} \models \bar{q}$ such that $F(\bar{a}, \bar{c})$,

$$S_\psi(\bar{a}, b) \text{ if and only if } S_\psi(\bar{c}, b).$$

Therefore there is a formula $\varphi_\psi(z) \in q(z)$ such that

$$\bar{q}(\bar{x}) \wedge \bar{q}(\bar{y}) \wedge F(\bar{x}, \bar{y}) \vdash \forall z(\varphi_\psi(z) \rightarrow (\rho_\psi(\bar{x}, z) \leftrightarrow \rho_\psi(\bar{y}, z))).$$

The formula

$$\forall z(\varphi_\psi(z) \rightarrow (\rho_\psi(\bar{x}, z) \leftrightarrow \rho_\psi(\bar{y}, z)))$$

defines an equivalence relation E_ψ. Note that $\bar{a}_{E_\psi} \in \mathrm{dcl}^{\mathrm{heq}}(\bar{a}_F)$ and it is in fact an imaginary.

CLAIM 3. $\bar{q}(\bar{x}) \wedge \bar{q}(\bar{y}) \wedge \bigwedge_{\psi \in \Sigma} E_\psi(\bar{x}, \bar{y}) \vdash E(x, y)$.

PROOF OF CLAIM 3. Assume $\bar{a} \models \bar{q}$ and $\bar{b} \models \bar{q}$. We check that $E_\psi(\bar{a}, \bar{b})$ implies $S_\psi(\bar{a}, b)$ and that $\bigwedge_{\psi \in \Sigma} S_\psi(\bar{a}, b)$ implies $E(a, b)$. The second statement is clear since on realizations of $\bar{q}$, $F(\bar{x}, \bar{y}) \vdash E(x, y)$ and since for each $\psi(x, y) \in \Sigma$ we may find some $\psi'(x, y) \in \Sigma$ such that $\psi'(x, z) \wedge \psi'(z, u) \wedge \psi'(u, y) \vdash \psi(x, y)$.

For the first statement it is enough to prove that $S_\psi(\bar{a}, a)$ whenever $\bar{a} \models \bar{q}$. To do so, choose $\bar{c} \models \bar{q}$ such that $F(\bar{c}, \bar{a})$ and $\bar{c} \perp_{\bar{a}_F} \bar{a}$. Then $E(c, a)$ and therefore $\models \psi_0(c, a)$. It suffices then to check that

$$\chi_0(x^-, \bar{c}) \wedge \psi(x, c) \wedge \psi(x, a) \text{ does not fork over } \bar{a}_F.$$

Let c^+ be the finite subtuple of $\bar{c}$ consisting of all its elements that do not occur in c. Then $\bar{c} = cc^+$ and $c \perp_{c_E} c^+$. Choose $d \equiv_{c^+} c$ such that $E(d, a)$ and

$d \mathop{\smile\!\!\!\!\!\!\mid}_{a_E c^+} \bar{a}\bar{c}$. Then, by conjugation, $d \mathop{\smile\!\!\!\!\!\!\mid}_{a_E} c^+$ and by transitivity $d \mathop{\smile\!\!\!\!\!\!\mid}_{a_E} \bar{a}\bar{c}$. Since $a_E \in \mathrm{dcl}^{\mathrm{heq}}(\bar{a}_F)$ and $\bar{a}_F \in \mathrm{dcl}^{\mathrm{heq}}(\bar{a})$, we get $d \mathop{\smile\!\!\!\!\!\!\mid}_{\bar{a}_F} \bar{a}\bar{c}$. Moreover $\models \chi_0(d^-, \bar{c}) \wedge \psi(d, c) \wedge \psi(d, a)$ because $\chi_0(x^-, \bar{c})$ is over c^+, $\models \chi_0(c^-, \bar{c})$, and $d_E = a_E = c_E$. ⊣

Now consider the sequence of imaginaries

$$\tilde{a} = (\bar{a}_{E_\psi} : \psi \in \Sigma).$$

Note that $a_E \in \mathrm{dcl}^{\mathrm{heq}}(\tilde{a})$ and $\tilde{a} \in \mathrm{dcl}^{\mathrm{heq}}(\bar{a}_F)$. By Lemma 19.20 $\mathrm{R}^{\mathrm{an}}(\tilde{a}/a_E) \leq \mathrm{R}^{\mathrm{an}}(\bar{a}_F/a_E)$ and we know that $\mathrm{R}^{\mathrm{an}}(\bar{a}_F/a_E) < \alpha$. Finally $\mathrm{SU}(\tilde{a}) < \infty$ since $\mathrm{SU}(\bar{a}) < \infty$. Thus $\tilde{a}$ satisfies all the required conditions and the proof of the lemma finishes. ⊣

REMARK 20.3. *A simplified version of Lemma* 20.2 *can be used to offer a different proof of Theorem* 18.29 *on elimination of bounded hyperimaginaries in low theories. Since the hyperimaginaries are bounded, no application of Lemma* 20.1 *is needed. Moreover the extension of a to $\bar{a}$ is not necessary since we can use type-definability of forking to justify easily the type-definability of $\overline{S_\psi}$. Hence it is enough to define S_ψ on realizations of $q(x)$ by: $S_\psi(a, b)$ if and only if there is some $c \models q$ such that*

- *$E(c, a)$ and $c \mathop{\smile\!\!\!\!\!\!\mid}_{a_E} b$,*
- *$\psi(x, c) \wedge \psi(x, b)$ does not fork over a_E.*

The hyperimaginary a_E turns out to be equivalent to the sequence of imaginaries $\tilde{a}$.

THEOREM 20.4. *Supersimple theories eliminate hyperimaginaries.*

PROOF. By Corollary 18.11 it suffices to show that T eliminates finitary hyperimaginaries. In fact we will prove a slightly stronger form of elimination: whenever a hyperimaginary a_0 is definable over an imaginary tuple a such that $\mathrm{SU}(a) < \infty$, then a_0 is eliminable.

The proof is by induction in $\mathrm{R}^{\mathrm{an}}(a/a_0)$. Assume that $\mathrm{R}^{\mathrm{an}}(a/a_0) = \alpha$ and the theorem holds for all a', a'_0 with $\mathrm{R}^{\mathrm{an}}(a'/a'_0) < \alpha$. Since $\mathrm{R}^{\mathrm{an}}(a/a_0) = \alpha$, for some $n < \omega$ there are hyperimaginaries $a_1, \ldots, a_n \in \mathrm{dcl}^{\mathrm{heq}}(a)$ and a sequence of ordinal numbers $(\alpha_i : i < n)$ such that $a \in \mathrm{bdd}(a_n)$, $a_i \in \mathrm{dcl}^{\mathrm{heq}}(a_{i+1})$, $\alpha_i \leq \alpha$, $\mathrm{tp}(a_{i+1}/a_i)$ is internal in the family of all formulas of D-rank α_i, and $\mathrm{tp}(a/a_i)$ is orthogonal to every type of D-rank $< \alpha_i$. By Lemma 18.6, a_n is equivalent to a sequence of imaginaries and we can replace it by this sequence and assume a_n is a sequence of imaginaries. Moreover $\mathrm{SU}(a_n) < \infty$ since $\mathrm{SU}(a) < \infty$ and a_n is definable over a. Note also that $a_0 \in \mathrm{dcl}^{\mathrm{heq}}(a_n)$ and $\mathrm{R}^{\mathrm{an}}(a_n/a_0) \leq \alpha$. We may proceed by a second induction on n. Note that $\mathrm{R}^{\mathrm{an}}(a_n/a_{n-1}) \leq \alpha_{n-1} \leq \alpha$. If $\alpha_{n-1} < \alpha$ we can use the general induction hypothesis to eliminate a_{n-1}. If $\alpha_{n-1} = \alpha$ we can use Lemma 20.2 to eliminate

a_{n-1} again. By the induction hypothesis on the length of the analysis, we finish. ⊣

COROLLARY 20.5. *Let T be supersimple. Then*

1. $a \overset{\text{Ls}}{\equiv}_A b$ *if and only if* $a \overset{\text{s}}{\equiv}_A b$.
2. *If* $A = \operatorname{acl}^{\text{eq}}(A)$, *then every* $p(x) \in S(A)$ *is an amalgamation base.*

PROOF. 1 is clear since by elimination of hyperimaginaries we have $\operatorname{Aut}(\mathfrak{C}/\operatorname{bdd}(A)) = \operatorname{Aut}(\mathfrak{C}/\operatorname{acl}^{\text{eq}}(A))$. For 2 notice that for any a, $\operatorname{tp}(a/A) \vdash \operatorname{tp}(a/\operatorname{bdd}(A))$. ⊣

DEFINITION 20.6. Let $p(x)$ be an amalgamation base. We say that a formula $\varphi(x, y) \in L$ is *p-stable* if $\mathfrak{p} \upharpoonright \varphi = \mathfrak{q} \upharpoonright \varphi$ for all $\mathfrak{p}, \mathfrak{q} \in \mathcal{P}_p$.

LEMMA 20.7. *Let T be simple and let $p(x)$ be an amalgamation base. If $\varphi(x, y) \in L$ is p-stable then for every $\mathfrak{p} \in \mathcal{P}_p$, $\mathfrak{p} \upharpoonright \varphi$ is definable.*

PROOF. We may assume $p(x) \in S(A)$ for a set A of real elements. By Corollary 5.23 there is a set of formulas over A, $\pi(x)$, such that for all a, $a \models \pi$ if and only if $p(x) \cup \{\varphi(x, a)\}$ does not fork over A, and similarly there is a set of formulas over A, $\pi'(x)$, such that all a, $a \models \pi'$ if and only if $p(x) \cup \{\neg\varphi(x, a)\}$ does not fork over A. By the uniqueness of the φ-type of any $\mathfrak{p} \in \mathcal{P}_p$, $\pi(x) \cup \pi'(x)$ is inconsistent. Choose $\psi(x) \in \pi(x)$ such that $\psi(x)$ is inconsistent with $\pi'(x)$. Then $\psi(x)$ is the definition we wanted. ⊣

PROPOSITION 20.8. *Let T be simple. A formula $\varphi(x, y) \in L$ is stable if and only if it is p-stable for every amalgamation base $p(x)$.*

PROOF. Assume $\varphi(x, y)$ is not p-stable for some amalgamation base $p(x) \in S(A)$. Then for some tuple a both $p(x) \cup \{\varphi(x, a)\}$ and $p(x) \cup \{\neg\varphi(x, a)\}$ do not fork over A. Let $(a_i : i < \omega)$ be a Morley sequence in $\operatorname{tp}(a/A)$. Since $p(x)$ is an amalgamation base and the sequence is A-independent, for every $X \subseteq \omega$,

$$p(x) \cup \{\varphi(x, a_i) : i \in X\} \cup \{\neg\varphi(x, a_i) : i \in \omega \setminus X\}$$

is consistent. Then $|S_\varphi(\{a_i : i < \omega\})| \geq 2^\omega$ and therefore $\varphi(x, y)$ is unstable.

Assume now that $\varphi(x, y)$ is unstable. There is an indiscernible sequence $(a_i b_i : i < \omega + \omega)$ such that $\models \varphi(a_i, b_j)$ if and only if $i < j$. Let $A = \{a_i b_i : i < \omega\}$. By Corollary 17.7 $p(x) = \operatorname{tp}(a_\omega/A)$ is an amalgamation base and by Lemma 10.4 $(a_i b_i : \omega \leq i < \omega + \omega)$ is a Morley sequence over A. Since $a_{\omega+2} \models \neg\varphi(x, b_{\omega+1})$ and $a_{\omega+2} \mathop{\smile\!\!\!\!\!\!\mid}_A b_{\omega+1}$, $p(x) \cup \{\neg\varphi(x, b_{\omega+1})\}$ does not fork over A. Since $a_\omega \models \varphi(x, b_{\omega+1})$ and $a_\omega \mathop{\smile\!\!\!\!\!\!\mid}_A b_{\omega+1}$, $p(x) \cup \{\varphi(x, b_{\omega+1})\}$ does not fork over A. This means that $\varphi(x, y)$ is not p-stable. ⊣

DEFINITION 20.9. Let $p(x)$ be an amalgamation base and let $\varphi(x, y) \in L$. We say that $\varphi(x, a)$ *is a canonical formula for* p if $\varphi(x, a) \in \mathfrak{p}$ for some $\mathfrak{p} \in \mathcal{P}_p$ and for every $a' \equiv a$, if $\varphi(x, a') \in \mathfrak{p}$ for some $\mathfrak{p} \in \mathcal{P}_p$ then $\varphi(x, a) \equiv \varphi(x, a')$.

THEOREM 20.10. *Let T be supersimple. For every amalgamation base $p(x)$, $\mathrm{Cb}(p)$ is equivalent to the set of canonical parameters of all canonical formulas for p.*

PROOF. By Theorem 20.4 we know that T eliminates hyperimaginaries and therefore $\mathrm{Cb}(p)$ is equivalent to a sequence of imaginaries c. Since $p \upharpoonright c$ is an amalgamation base, we may assume $p(x) \in S(c)$. It is clear that all canonical parameters of canonical formulas for p are definable over c. By supersimplicity there is a finite subtuple c_0 of c such that p does not fork over c_0. By Lemma 17.23, $c \in \mathrm{bdd}(c_0)$ and since they are tuples of imaginaries, $c \subseteq \mathrm{acl}^{\mathrm{eq}}(c_0)$.

We show now that every finite subtuple c_1 of c extending c_0 is a canonical parameter of some canonical formula for p. For any tuple d of the length of c we let d_1 be the corresponding finite subtuple of d. We write $p(x) = p(x, c)$ with $p(x, z) \in S(\emptyset)$. Notice that for any $c' \equiv c$,

$$c = c' \text{ if and only if } p(x, c) \cup p(x, c') \text{ does not fork over } c.$$

Hence, if $c' \equiv c$ and $c'_1 \neq c_1$, then $p(x, c) \cup p(x, c')$ forks over c. By Proposition 5.22 and Remark 3.14 there are formulas $\varphi(x, y), \psi(x, z_2) \in L$, a finite subtuple c_2 of c extending c_1, and a natural number k such that for all $c'_2 \equiv c_2$,

$$c_1 \neq c'_1 \text{ implies } \mathrm{D}(\psi(x, c_2) \wedge \psi(x, c'_2), \varphi, k) < \mathrm{D}(p(x, c), \varphi, k).$$

Let $n = \mathrm{D}(p(x, c), \varphi, k)$. Extending $\psi(x, z_2)$ and c_2 if necessary, we may assume that $\mathrm{D}(\psi(x, c_2), \varphi, k) = n$. Since c_2 is algebraic over c_1, there is a formula $\chi(z_2, z) \in L$ such that $\models \chi(c_2, c_1)$ and $\chi(z_2, c_1) \vdash \mathrm{tp}(c_2/c_1)$. Let

$$\psi'(x, z) = \exists z_2(\psi(x, z_2) \wedge \chi(z_2, z)).$$

Since c_2 is algebraic over c_1, $\psi'(x, c_1)$ is equivalent to a finite disjunction of c_1-conjugates of $\psi(x, c_2)$ and then, by Proposition 3.12, $\mathrm{D}(\psi'(x, c_1), \varphi, k) = n$. For the same reason, for all $c'_1 \equiv c_1$,

$$c'_1 \neq c_1 \text{ implies } \mathrm{D}(\psi'(x, c'_1) \wedge \psi'(x, c_1), \varphi, k) < n.$$

It follows that $\psi'(x, c_1)$ is a canonical formula for $p(x)$ and that c_1 is its canonical parameter. ⊣

PROPOSITION 20.11. *Let T be simple and let $p(x)$ be an amalgamation base. If $\mathrm{Cb}(p)$ is equivalent to the set of canonical parameters of all canonical formulas for p, then $\mathrm{Cb}(p)$ is also equivalent to the set of canonical parameters of the definitions of all $\mathfrak{p} \upharpoonright \varphi$ for all $\mathfrak{p} \in \mathcal{P}_p$ for all p-stable formulas $\varphi(x, y)$.*

PROOF. The canonical base of p is equivalent to a tuple of imaginaries c. We may assume $p(x) \in S(c)$. For any p-stable $\varphi(x, y) \in L$ we know, by Lemma 20.7, that the common φ-type of any $\mathfrak{p} \in \mathcal{P}_p$ is definable. Let c_φ be the canonical parameter of such a definition. It is clear that each c_φ is definable over c. Now we show that each canonical parameter d of each canonical formula for p is in fact the canonical parameter c_φ of some p-stable formula $\varphi(x, y)$. Let $\psi(x, y) \in L$ and assume that $\psi(x, d)$ is canonical for p with canonical parameter d. Let $r(y) = \text{tp}(d)$. If $d' \models r(y)$ and $d \neq d'$, then $\psi(x, d) \not\equiv \psi(x, d')$ and therefore $p(x) \cup \{\psi(x, d')\}$ forks over c. By Corollary 5.23, there is some formula $\theta(y) \in r(y)$ such that for all $d' \models \theta(y)$, if $d \neq d'$, then $p(x) \cup \{\psi(x, d')\}$ forks over c. Let $\varphi(x, y) = \psi(x, y) \wedge \theta(y)$. Then $\varphi(x, d') \in \mathfrak{p}$ for some $\mathfrak{p} \in \mathcal{P}_p$ if and only if $d = d'$. The only positive instance of $\varphi(x, y)$ in a type in the amalgamation class $\mathcal{P}_p$ is $\varphi(x, d)$. Note that $\neg\varphi(x, d)$ does not appear in any such type since it is a formula over c and all types in $\mathcal{P}_p$ are extensions of p. This implies that $\varphi(x, y)$ is p-stable and $d \sim c_\varphi$. ⊣

COROLLARY 20.12. *Let T be supersimple. For every amalgamation base $p(x)$, $\text{Cb}(p)$ is equivalent to the sequence of all canonical parameters of the definitions of all $\mathfrak{p} \restriction \varphi$ for all $\mathfrak{p} \in \mathcal{P}_p$ for all p-stable formulas $\varphi(x, y)$.*

PROOF. By Theorem 20.10 and Proposition 20.11. ⊣

REFERENCES

[1] Hans Adler, *Strict orders prohibit elimination of hyperimaginaries*, Preprint, February 2007.

[2] ———, *A geometric introduction to forking and thorn-forking*, ***Journal of Mathematical Logic***, vol. 9 (2009), no. 1, pp. 1–20.

[3] ———, *Thorn-forking as local forking*, ***Journal of Mathematical Logic***, vol. 9 (2009), no. 1, pp. 21–38.

[4] John T. Baldwin, ***Fundamentals of Stability Theory***, Springer-Verlag, Berlin, 1988.

[5] Itay Ben-Yaacov, Anand Pillay, and Evguenii Vassiliev, *Lovely pairs of models*, ***Annals of Pure and Applied Logic***, vol. 122 (2003), pp. 235–261.

[6] Steven Buechler, *Lascar strong types in some simple theories*, ***The Journal of Symbolic Logic***, vol. 64 (1999), pp. 817–824.

[7] Steven Buechler, Anand Pillay, and Frank O. Wagner, *Supersimple theories*, ***Journal of the American Mathematical Society***, vol. 14 (2000), pp. 109–124.

[8] Enrique Casanovas, *Simplicity simplified*, ***Revista Colombiana de Matemáticas***, vol. 41 (2007), pp. 263–277.

[9] Enrique Casanovas and Byunghan Kim, *A supersimple nonlow theory*, ***Notre Dame Journal of Formal Logic***, vol. 39 (1998), pp. 507–518.

[10] Enrique Casanovas, Daniel Lascar, Anand Pillay, and Martin Ziegler, *Galois groups of first order theories*, ***Journal of Mathematical Logic***, vol. 1 (2001), pp. 305–319.

[11] Enrique Casanovas and Frank O. Wagner, *The free roots of the complete graph*, ***Proceedings of the American Mathematical Society***, vol. 132 (2004), no. 5, pp. 1543–1548.

[12] Bradd Hart, Byunghan Kim, and Anand Pillay, *Coordinatisation and canonical bases in simple theories*, ***The Journal of Symbolic Logic***, vol. 65 (2000), pp. 293–309.

[13] Wilfrid Hodges, ***A Shorter Model Theory***, Cambridge University Press, Cambridge, 1997.

[14] Ehud Hrushovski, *Kueker's conjecture for stable theories*, ***The Journal of Symbolic Logic***, vol. 54 (1989), pp. 207–220.

[15] BYUNGHAN KIM, ***Simple First Order Theories***, Ph.D. thesis, University of Notre Dame, June 1996.

[16] ———, *Forking in simple unstable theories*, ***Journal of the London Mathematical Society***, vol. 57 (1998), pp. 257–267.

[17] ———, *A note on Lascar strong types in simple theories*, ***The Journal of Symbolic Logic***, vol. 63 (1998), pp. 926–936.

[18] ———, *Simplicity, and stability in there*, ***The Journal of Symbolic Logic***, vol. 66 (2001), pp. 822–836.

[19] BYUNGHAN KIM and ANAND PILLAY, *Simple theories*, ***Annals of Pure and Applied Logic***, vol. 88 (1997), pp. 149–164.

[20] ———, *From stability to simplicity*, ***The Bulletin of Symbolic Logic***, vol. 4 (1998), pp. 17–36.

[21] ———, *Around stable forking*, ***Fundamenta Mathematicae***, vol. 170 (2001), pp. 107–118.

[22] DANIEL LASCAR, *On the category of models of a complete theory*, ***The Journal of Symbolic Logic***, vol. 47 (1982), pp. 249–266.

[23] ———, ***Stability in Model Theory***, Longman Scientific & Technical, Harlow, U.K., 1987.

[24] ———, ***La théorie des modèles en peu de maux***, Nouvelle bibliothèque mathématique, no. 10, Cassini, Paris, 2009.

[25] DANIEL LASCAR and ANAND PILLAY, *Forking and fundamental order in simple theories*, ***The Journal of Symbolic Logic***, vol. 64 (1999), pp. 1155–1158.

[26] ———, *Hyperimaginaries and automorphism groups*, ***The Journal of Symbolic Logic***, vol. 66 (2001), pp. 127–143.

[27] MIHÁLY MAKKAI, *A survey of basic stability theory with particular emphasis on orthogonality and regular types*, ***Israel Journal of Mathematics***, vol. 49 (1984), pp. 181–238.

[28] DAVID MARKER, ***Model Theory: An Introduction***, Graduate Texts in Mathematics, vol. 217, Springer, New York, 2002.

[29] LUDOMIR NEWELSKI, *The diameter of a Lascar strong type*, ***Fundamenta Mathematicae***, vol. 176 (2003), no. 2, pp. 157–170.

[30] LUDOMIR NEWELSKI and MARCIN PETRYKOWSKI, *Weak generic types and covering of groups I*, ***Fundamenta Mathematicae***, vol. 191 (2006), pp. 201–225.

[31] RODRIGO PELÁEZ, ***About the Lascar Group***, Ph.D. thesis, University of Barcelona, April 2008.

[32] ANAND PILLAY, *Forking, normalization and canonical bases*, ***Annals of Pure and Applied Logic***, vol. 32 (1986), pp. 61–81.

[33] ———, ***Geometric Stability Theory***, Oxford University Press, 1996.

[34] ———, *Definability and definable groups in simple theories*, ***The Journal of Symbolic Logic***, vol. 63 (1998), pp. 788–796.

[35] ANAND PILLAY and BRUNO POIZAT, *Pas d'imaginaires dans l'infini!*, ***The Journal of Symbolic Logic***, vol. 52 (1987), no. 2, pp. 400–403.

[36] BRUNO POIZAT, ***Cours de théorie des modèles***, Nur al-Mantiq wal-Ma'rifah, 82, rue Racine 69100 Villeurbanne, France, 1985, Diffusé par OFFILIB.

[37] ZIV SHAMI, *Definability in low simple theories*, ***The Journal of Symbolic Logic***, vol. 65 (2000), pp. 1481–1490.

[38] SAHARON SHELAH, *Simple unstable theories*, ***Annals of Mathematical Logic***, vol. 19 (1980), pp. 177–203.

[39] ———, ***Classification Theory***, second ed., North Holland P.C., Amsterdam, 1990.

[40] KATRIN TENT and MARTIN ZIEGLER, ***A Course in Model Theory***, Lecture Notes in Logic, Cambridge University Press, Cambridge, to appear.

[41] FRANK O. WAGNER, ***Simple Theories***, Mathematics and Its Applications, vol. 503, Kluwer Academic Publishers, Dordrecht, 2000.

[42] MARTIN ZIEGLER, *Stabilitätstheorie*, Freiburger Vorlesung gehalten im Wintersemester 1988/1989. Ausgearbeitet von Urs Künzi, Oktober 1991.

INDEX